The Idea of the Sciences in the French Enlightenment

The Idea of the Sciences in the French Enlightenment

A Reinterpretation

G. Matthew Adkins

UNIVERSITY OF DELAWARE PRESS
Newark

University of Delaware Press

Printed in the United States of America on acid-free paper
Distributed by the University of Virginia Press

ISBN 978-1-64453-064-1 (paper)
ISBN 978-1-64453-065-8 (ebook)

British Library Cataloguing-in-Publication Information Available

Library of Congress Cataloging-in-Publication Data

Adkins, G. Matthew (Gregory Matthew), 1971–
The idea of the sciences in the French enlightenment : a reinterpretation / G. Matthew Adkins.
pages cm
Includes bibliographical references and index.
1. Science—France—History—17th century. 2. Science—France—History—18th century.
3. Science, Renaissance. 4. Enlightenment—France. I. Title.
Q127.F8A35 2014
509.44'09033—dc23
2013035178

For Miriamne

Contents

Acknowledgments

This book has been a long time coming and has incurred in me, its author, a number of debts to those who made it possible for me to complete it. I am very grateful for the opportunity here to acknowledge at least some of those debts and to offer heartfelt thanks to those to whom these debts are owed, though they are never truly repayable, except perhaps in the best possible coinage of gratitude, friendship, and offers of mutual obligation.

Years ago this project first began as a doctoral dissertation at the University of North Carolina at Chapel Hill. Although reconceived and rewritten (multiple times) in the intervening years, I would like to thank those who first helped me to develop the ideas presented here, including John Headley, Michael McVaugh, Lloyd Kramer, Don Reid, and Jay Smith. I also offer special thanks to John Chasteen, who ran a stimulating dissertation design course in which I first began to write down the ideas that eventually became this book. His seminar was an unforgettable experience, during which I wrote the first version of what is chapter 2 of the present book, originally published in the *Journal of the History of Ideas*—and still my personal favorite chapter of this work.

Of those with whom I worked and studied at Chapel Hill, however, the greatest thanks must go to Jay Smith. Over the years since I left Chapel Hill to take up an itinerant professorial life that only relatively recently settled down, Jay has remained an invaluable friend and advisor in matters both professional and scholarly. I am especially grateful for his intensive reading of the final version of this manuscript before I submitted it to the University of Delaware Press. His revision suggestions were critical to the clarification and presentation of this book's argument.

Professor Donald Mell at the University of Delaware Press supported the publication of this book. Two anonymous readers for the press not only recommended publication but also offered extremely helpful and very detailed final revision suggestions. I owe them all deep thanks for their commitment to promoting and sustaining new scholarship and the exchange of ideas.

Over the years, I have held several academic positions, each coming with significant teaching obligations that I did not neglect. As a result, I have devoted such time as I was able to spare from teaching to the writing, rewriting, rewriting again, and near-final editing of this book. Such time translated into isolated moments of revision during the early mornings and weekends, and during short summers when I tried to lose myself in scholarship, writing, and thinking. I therefore owe an unrepayable debt to my parents, Greg and Marguerite Adkins, for understanding why I was so lost in my work; to my wife's parents, professors Bill and Rifke Krummel, for their unfailing support; and to my beautiful daughters, Yetta and Shoshana, for the necessary distractions and occasional headaches—always worth every moment of time I gave to them. Most of all, I thank my partner, Miriamne Krummel, for understanding exactly how I feel about my scholarly interests, for she feels the same way about her own scholarship. I dedicate this book to her.

Parts of chapter 1, "The Montmor Discourse: Samuel Sorbière and the Foundation of the Royal Academy of Sciences," originally appeared in different form as "The Montmor Discourse: Science and the Ideology of Stability in Old Régime France," in the *Journal of the Historical Society* 5, no. 1 (March 2005): 1–28. Parts of chapter 2, "The *Esprits Supérieurs*: Bernard de Fontenelle's Academic Eulogies," originally appeared in different form as "When Ideas Matter: The Moral Philosophy of Fontenelle," in the *Journal of the History of Ideas* 61, no. 3 (July 2000): 433–52. John Wiley & Sons, the publisher of the *Journal of the Historical Society*, and the University of Pennsylvania Press, the publisher of the *Journal of the History of Ideas*, have kindly permitted republication of that material.

Introduction

From the mid-seventeenth century until the end of the eighteenth century, some French savants developed and tried to implement politically an idea of the natural sciences as morally enlightening to those who pursued them. During that time, France was experiencing a series of interconnected social, economic, political, military, and even cultural crises that would eventually spiral out of control and lead to the Revolution of 1789.[1] The idea that the natural sciences provided moral as well as intellectual enlightenment, of course, implied that those who devoted their time to the study of the sciences became morally superior members of society. In an era of prolonged and despair-inducing crises that seemed to affect all aspects of French society, that special idea of the natural sciences led those savants most attached to the idea to the conclusion—already well understood, since it was, of course, not really a conclusion at all, but an assumption—that savants had, or should have, a special social responsibility to effect desperately needed reforms suggested to them by their study of the natural sciences. In short, the idea of the natural sciences promoted the importance of savants (at once experts, scholars, scientists, and intellectuals) in the remedying of social—and political—ills.

In this book I trace the development of the idea of the sciences as morally enlightening—as well as the unexpected conflicts to which it gave rise—for the deployment of the idea in the public arena led more often to frustration than to success. Savants in or closely associated with the French Royal Academy of Sciences (*Académie royale des sciences*) articulated and deployed the idea over the course of the mid-seventeenth to the late eighteenth centuries, and its implications grew more radical as the crises of the eighteenth century worsened. I try to show how and why this radicalization—and eventually the emergence of revolutionary sentiments—occurred. The story I tell here is about the rise of one of the central features of the *narrative* of the Enlightenment—that is, the story that late seventeenth- and early eighteenth-century French thinkers created to assert the existence of their enlightened era (an assertion quite often flying in the face of most observable facts). In a recent intellectual history (or what he prefers to term a *geneology*) of the Enlightenment, Dan Edelstein puts the narrative concisely: according to French writers of the era, "the present age (*siècle*) was 'enlightened' (*éclairé*) because the 'philosophical spirit' of the Scientific Revolution had spread to the educated classes, institutions of learning, and even parts of the government. More generally, they argued, changes in science had led to changes in society."[2] The story I tell here climbs down into the nitty-gritty details to ask: What was this idea of the sciences (this "philosophical spirit")? How and why did it emerge? And what happened to the idea as the crises of France worsened and precipitated a revolution?

Although this book is a story of an idea, ultimately the story is about people who developed and then deployed the idea for many reasons: their motives (so much as we can guess or suppose them from their actions and writings) seem to have been both personal and social, selfish and altruistic, ethical and practical—undoubtedly, like most human motives, never purely one thing or another, or even necessarily understood fully by those who held them, or wholly understandable by anyone except those who held them. This does not make their ideas bad ideas, and even if some of us now might have a more cynical view of the sciences, I am not sure that is a good reason to impose our cynicism on their less cynical age.

I do not, however, present the moral idea of the natural sciences as merely the tool of people operating with free will or autonomy. Despite an interesting and vigorous debate regarding how to recover methods of representing human "agency" from the linguistic and social determinisms that have dominated historiography for a very long time, I do not think that agency is ever discretely held in the power of an autonomous will but is distributed around and through individuals.[3] Historians of the French Revolution and its intellectual origins in particular have long participated in the debate over agency, since explaining why people might opt for dramatic social and political changes seems to require an adequate way to theorize and represent human agency.[4] In a broader sense, if our ideas do not matter, then in a real way human intentionality does not matter, either. The predominance of discourse analysis in French historiography, however, introduced a noted problem that Michael Fitzhugh and William Leckie highlighted in a 2001 article in *History and Theory* proclaiming the impossibility of conceiving of human agency while yet maintaining the dominant epistemology of the linguistic turn.[5] "Without recourse to metaphysics," Fitzhugh and Leckie wrote, "propositions about diachronic linguistic change and humans who cognize entirely with or in a closed system are ultimately incompatible; these principles require that people cause diachronic change even when beings locked inside the system have no way of effecting such change."[6] Fitzhugh and Leckie proposed that historians turn to cognitive science and neurology (a "cognitive turn") when scientists have already shown that humans engage in a significant amount of cognition prior to or outside of language—thus demonstrating that language is not, in fact, a closed system at all and that perturbations arrive to us from the "great outdoors."[7] If humans are able to mediate experiences that transcend discursive constructions, then agency and change become a possibility. In the same issue of *History and Theory*, however, Jay Smith entered the debate with a warning that we cannot suppose the ontological priority of "experience," arguing that raw experience is always mediated by dynamic "interpretive dispositions" on the part of the subject.[8]

Over the past decade both continental philosophy and sociology have converged with and offered nuance and complication to this historiographical debate. One of the implications of speculative realism, like its older cousin, Bruno Latour's actor-network theory (ANT), as well as its offspring, object-oriented ontology, is a flat ontology in which all objects (i.e., "ideas," "bodies," "oranges," and even fictional entities) possess the same ontological weight and are all actants in the sense that they have the ability to make a difference to other objects that possess the power to enter into relations with them.[9] According to these models, individual people exercise (or seem to exercise) agency, but not just as they please—and not because they are trapped within enveloping structures (such as language or power relations) that do the actual acting—but rather because people operate in a highly complex field of associations, the borders of which they (or we) might not even grasp clearly.[10] Agency, as I try to present it here, drifts around or is always "dislocated, articulated, delegated, translated" within a field of associations.[11] The idea of the natural sciences, once articulated and deployed, became yet another actant in the partial field of associations I will attempt to trace or account for here, exercising a form of agency, manipulating or being manipulated by the people upon whom I focus this book.[12] Or possibly the idea of the natural sciences as "yet another" actant is too simplistic, for, of course, the idea itself was not some sort of free-floating entity communicating itself but was contained and mediated by the human and nonhuman objects that themselves formed or constituted the field of associations through those acts of mediation. It is, perhaps, the idea that allows me to see the associations; the idea is that which gets translated from one actor to another. Was the idea mediated through oral language, inflected by the position and rhetoric of the speaker? Is the idea contained in a physical product of labor—a letter, a book, an essay, or a government report? All of these different objects matter; they form the material existence of the idea that, though constantly changing (perhaps evolving), we can, nevertheless, somehow recognize as a single idea acting upon those who found the idea compelling or perturbing just as much as the idea was itself adapted and translated by them according to the dynamics of their desires, needs, and circumstances. None of these mediations was entirely "free." For example, the marquis de Condorcet in the last third of the eighteenth century was compelled to deal with the idea of the sciences because of the associations in which he found himself embroiled through the historical process of his life. But nor are such mediations "determined" by anything outside or behind the field of associations, although tracing their boundaries might take us on an impossibly long tour of the cognitive process.

For, of course, the idea of the sciences as morally enlightening was not exactly original in itself in the seventeenth century. The men (indeed, they are all men—there were no women in the Royal Academy of Sciences) whose stories I tell here did not suddenly come up with an entirely new idea, hitherto unknown to the world. They formed their idea of the sciences from within a context of philosophical thinking dating at least back to Plato—as anyone who has read Plato's famous *Republic* knows quite well.[13] The most immediate context for the reactivation of the idea, however, was the Neostoic philosophical discourse playing itself out in France from the mid-sixteenth through the seventeenth centuries. As Edelstein writes, "[T]heirs was just as much a story about antiquity as it was about modernity." Although I leave the otherwise important quarrel of the ancients and moderns on the sidelines of my story, the power of ancient literature and ancient forms of thought over early modern thinkers is such that they form an essential part of any narrative of the Enlightenment. In this story, I stick closely to the influence of very specific forms of ancient thought: Platonic, Aristotelian, and especially Stoic.

Historians have long argued that early modern thinkers revived the ancient Greco-Roman philosophy of Stoicism first in the context of the Dutch Revolt (1568–1648) and then in France during the Wars of Religion (1562–1598) as a response to the growing sense of social and political crisis and disorder that many sensible people felt at the time.[14] Further, the revivers of Stoicism considered the philosophy not just a means to personal tranquility (the term often used in place of the ancient Greek *apatheia*—the other common Neostoic term being *constancy*), but as a key to resolve larger social and political crises. Although historians have challenged the old notion of a "general crisis" of the seventeenth century that precipitated the birth of "modernity," crisis remains a very useful concept for understanding the way experiences were articulated in the era.[15] Regardless of whether there was a real, objective "general crisis," there certainly were many interconnected and significant events—themselves dynamics involving multiple variables, interpretations, and actions—that threatened the perceived stability of early modern political, religious, cultural, and social modalities. These events gave rise to a sense of crisis articulated quite clearly by the classically educated elite, especially in France—a group that was already predisposed to deploy the trope of "crisis" to make sense of changes, especially disorienting ones.[16] And certainly a significant publishing section of educated elite European society interpreted the disorienting changes that arose in the wake of the voyages of discovery, the Protestant Reformation, and the violent confessional wars of the sixteenth and early seventeenth centuries as a crisis of order. A perfectly tranquil social, political, and cultural order might never have existed—it obviously did not—but many people interpreted the intensity of change in the early modern era as a loss of order. The result of that interpretation first arrived, as outlined lucidly in Gerhard Oestreich's now classic but controversial study, in the Dutch Republic with the scholar Justus Lipsius's revival of Roman Stoicism.[17] With Lipsius and his followers, especially in the Netherlands and France, the desire for order prompted a reexamination of practical Roman Stoic thought and its application to the political, military, social, and intellectual spheres as a source of discipline and constancy meant to reestablish the grounds for order and tranquility.

The resulting Neostoic philosophical discourse introduced the possibility of thinking not only rationally but also scientifically about social and political order—an implication that would have considerable impact on the development of Enlightenment thought in the eighteenth century—and on the development of the idea of the morally enlightening natural sciences that concerns us here. Even from Lipsius's example, however (he ultimately sided with the Spanish during the Dutch Revolt), the Dutch and later French variants of Neostoicism linked the possibility of establishing social and political discipline to the existence of an enlightened, but absolute, ruler. Although the Netherlands, particularly with Baruch Spinoza's circle, played an important role in the development of "radical" pro-republican Enlightenment thought, another far more influential Dutch trend (at least in the short term) favored the strengthening of monarchical power through Stoic disciplinary reforms as the key to finding a cure for the ills of society.[18] In its origins, the Neostoic goal held close to the curve of Platonic thought: the hope for order, tranquility, and even justice in society lay in the hands of intellectual superiors who would prepare the way, if not for a philosopher-king, then for an enlightened despotism.

By the late seventeenth century, many savants and other thinkers in France were busy criticizing the foolishness of their philosophical predecessors and excoriating the supposed antiemotionalism of Stoicism as ridiculous and impossible, but they continued to operate within the boundaries of Neostoic philosophical discourse, and they still sought to accomplish generally Neostoic goals. The idea of the morally enlightening natural sciences stemmed from an increased confidence (or was it desperate hope?) that the natural sciences might allow a savant to discern clearly the way things actually were—to perceive the ultimate ontological reality, or at the very least to discern the way things most *probably* were. Since Pierre Gassendi, for one, championed the idea that our knowledge of the natural world was only probable, not certain, probability became a way in that era to avoid the stain of dogmatic thought, to sustain a certain intellectual modesty—to say "I'm not trying to stop the inquiry into nature by insisting that I know everything already."[19] The savant who pursued the natural sciences properly—with the discipline necessary to discover the (probable) truth about the world—achieved moral superiority because he could direct his natural and undeniable human passions toward that truth. Although obviously derived from Platonism, the notion that one could achieve moral superiority (constancy, tranquility, clarity, and discipline) by understanding the true nature of reality was the quintessential Stoic ideal. This ancient ideal, however, formed the basic pattern of intellectual modernity: most benighted members of society, according to this ideal, live behind a veil of false consciousness, the illusions of our social fabric—religion, culture, convention, fashion, and more. Society might get along just fine so long as nothing goes fundamentally wrong with the reality of our social structures (whatever those might be). But when an emergency does exist, a need for correction or reform, or a restitching of a ripped social fabric, the best possible reformers are the scientists who see behind the veil, who know what is really—or at least most probably—going on.[20]

In the seventeenth century, no one (or no one among the people followed here) considered pulling down the veil altogether. The inferior minds that made up the majority of society needed their illusions, and the superior minds of the savants openly but cynically followed social conventions—such as Catholic orthodoxy in France—while working behind the scenes to put society on a more rational footing. Only in the eighteenth century did the idea of the morally enlightening natural sciences find itself transformed into the Enlightenment ideal of emancipation through education, in which false consciousness became the weapon of the enemies of humanity—a weapon the defenders of light sought to break in order to free society from the tyranny of darkness, superstition, and corruption.

But that is the story I wish to tell here, so rather than further belaboring the introduction, I now begin the story more or less in media res: I start in April 1663, on the eve of the creation of the Royal Academy of Sciences, and end in March 1794, upon the death of the marquis de Condorcet, the last perpetual secretary of the Academy before the Revolutionary government disbanded it in August 1793. In those 131 years, one of the most long-lasting and important ideals of the Enlightenment came into being and continues to inform our contemporary society. It is important to remember that the moral idea of the sciences formed in an atmosphere in which the people about whom I write sensed a profound crisis in their society—a feeling of crisis that led many of them to despair. Their writings reveal their experience of crisis clearly: to the account below, therefore, I will try to add little or nothing that was not already there. I forward no social causal theory here (e.g., capitalism, class, culture, discourse, etc.) that is not already articulated by the actors. While I do hope to make a critical and historiographical intervention into our understanding of the Enlightenment—or one particular, and even originary, strand of what historians now see as the large and diverse Enlightenment—my goal is not to debunk the expressed motives and drives of the people here described by revealing what was really going on. The savants in this story can, and do, express themselves quite clearly.

NOTES

1. Citations of published material are shortened in the endnotes. Full bibliographical information is available in the bibliography.

The literature on the problems and crises of the Old Regime would be far too great to list, but a recent volume of essays that takes stock of postrevisionist accounts of the origin of the Revolution is Thomas Kaiser and Dale Van Kley, eds., *From Deficit to Deluge: The Origins of the French Revolution*.

2. Dan Edelstein, *The Enlightenment: A Geneology*, 8.

3. See Jay M. Smith, "Between *Discourse* and *Experience*"; Michael Fitzhugh and William Leckie Jr., "Agency, Postmodernism, and the Causes of Change"; Bruno Latour, *Reassembling the Social*; and especially Julian Yates, "Toward a Theory of Agentive Drift; Or, a Particular Fondness for Oranges circa 1597."

4. See Jay Smith, "Between *Discourse* and *Experience*," and Timothy Tackett, *Becoming a Revolutionary*.

5. Keith Baker also sought to make an end run around the problem of agency in his defense of discourse analysis in *Inventing the French Revolution*, 6–7.

6. Fitzhugh and Leckie, "Agency, Postmodernism, and the Causes of Change," 74. These authors (among others) argue powerfully that the philosophical premises underlying the linguistic turn—which has exercised considerable influence in cultural history since the 1980s—require the logical denial of agency as a causal force in history.

7. Fitzhugh and Leckie, "Agency, Postmodernism," 78. Of particular interest is research cited by the authors that argues for the "decentering" of our understanding of human cognition in that "distinctions between mind, body, and world need to be abandoned." Quentin Meillassoux uses the phrase *great outdoors* to refer to the problem of "correlationist" philosophies, of which poststructuralism is the ultimate expression, in which we are locked inside language or consciousness without being able to get out, and it becomes impossible to conceive of humanity as anything but an effect of the structures that contain it (thus denying the possibility of human agency). See Quentin Meillassoux, *After Finitude: An Essay on the Necessity of Contingency*, 4–6.

8. Smith, "Between *Discourse* and *Experience*," 116–42.

9. See Levi Bryant et al., eds., *The Speculative Turn*; Bruno Latour, *Reassembling the Social*; Julian Yates, "Toward a Theory of Agentive Drift"; and Jane Bennett, "Powers of the Hoard: Further Notes on Material Agency." As Bennett writes on page 257, "[T]he 'I,' as a compound of human and non-human parts, is continually entering and leaving larger assemblages (ideologies, diets, cultures, technological regimes) made up of other sets of composite or compound bodies. A full acknowledgment of the porosity and contagion between bodies would entail a dramatic revision of the role of 'will' and 'intentionality' in human agency."

10. "Men make their own history, but they do not make it just as they please; they do not make it under circumstances chosen by themselves, but under circumstances directly encountered, given and transmitted from the past. The tradition of all the dead generations weighs like a nightmare on the brain of the living." Karl Marx, *The Eighteenth Brumaire of Louis Bonaparte*, 15.

11. Or what ANT theorists call an "actor-network." Bruno Latour, *Reassembling the Social*, 166.

12. See Bruno Latour, *Reassembling the Social.*

13. See, in particular, the Allegory of the Cave, *The Republic*, 514a–521d.

14. See, in particular, Gerhard Oestreich, *Neostoicism and the Early Modern State*, and Nannerl Keohane, *Philosophy and the State in France: The Renaissance to the Enlightenment*, which both argue this point. I would also direct readers to Christopher Brooke, *Philosophic Pride: Stoicism and Political Thought from Lipsius to Rousseau*, which unfortunately appeared after I had completed the major drafts of this book (in 2012).

15. Paul Hazard's idea of a "crisis of the European mind" might obtain new usefulness with the recognition of the way early modern thinkers articulated their experiences of the changes they lived through. See especially the fascinating *American Historical Review* Forum, "The General Crisis of the Seventeenth Century Revisited," in the October 2008 issue of *American Historical Review* 113, no. 4 (October 2008): 1029–99.

16. See J. B. Shank, "Crisis: A Useful Category of Post-Social Scientific Historical Analysis?" *American Historical Review* 113, no. 4 (October 2008): 1090–99. Shank notes that the notion of crisis was first introduced from ancient Greek medical thought into historical thought by Thucydides.

17. Justus Lipsius, *De Constantia Libri Duo, qui alloquium praecipue continent in publicis malis* (1584). Some scholars have recently and significantly challenged both the Stoicism of Lipsius's philosophy and the historiographical agenda of Gerhard Oestreich. See Brooke, *Philosophic Pride*, 15–16.

18. Jonathan Israel, *Radical Enlightenment.*

19. See, for example, Pierre Gassendi, *Exercitationes paradoxicae adversus Aristoteleos* (1624).

20. See Bruno Latour's discussion of this aspect of modern social science in *Reassembling the Social*, 48, and throughout.

Chapter One

The Montmor Discourse: Samuel Sorbière and the Foundation of the Royal Academy of Sciences

On April 3, 1663, a Tuesday, a savant and physician named Samuel Sorbière (1615–1670) pronounced a discourse on the natural sciences at a meeting of savants in the Parisian home of Henri-Louis Habert de Montmor, who regularly hosted a sort of informal scientific academy. History has largely forgotten Sorbière's speech. Those who do remember it—generally historians of early modern science—consider it primarily of symbolic importance, for it contained one of the first serious proposals in France for a royally sponsored academy of the sciences.[1] Samuel Sorbière's speech was, according to general accounts, merely a blip in the development of modern state-sponsored science, barely worth notice except as a curiosity. Someone had to stand up and make a proposal, and history assigned this small honor to an obscure savant who, after his short moment as a world-historic actor, soon withdrew behind the curtains of time and disappeared. But when viewed in a wider sweep of history, Sorbière's speech proves more than a mere curiosity. In his discourse, Sorbière voiced a new conception of the relationship between the sciences and society—and a new role for savants, the makers of scientific knowledge. His speech marked a new beginning in Neostoic philosophical thought—in fact, the speech helped to initiate an Enlightenment ideal that sought to apply the language of the natural sciences to the sociopolitical realm in order to make the world conform to Neostoic ideals such as rationality, harmony, and stability.

Since the appearance of Charles Gillispie's *Science and Polity in France at the End of the Old Regime* (1980), the preponderance of research has in fact rejected Gillispie's assertion that the statesmen and politicians of the Old

Regime sought in the sciences only instrumental powers—"weapons, techniques, information, communications, and so on"—and that scientists wanted only to avoid politicization.[2] The opposite seems to be true: the sciences became a source, albeit contested, of political authority and legitimation.[3] Keith Baker, in his landmark essay "On the Problem of the Ideological Origins of the French Revolution," identifies a discursive triad that defined the rhetoric of French monarchical authority during the Old Regime: the discourses of justice, will, and reason.[4] The political discourse of reason, Baker argues, also provided the rhetoric of modernity by which the economist and finance minister, Robert-Jacques Turgot, and his fellow physiocrats in the 1760s and 1770s wanted "to transpose the problem of social order into the language of social science"—in short, to transform royal will into the exercise of universal reason, and so make any opposition to their reforms seem tantamount to denying the dictates of that reason. (I return to the subject of Turgot's reforms in chapter 3.) Regardless of whether we accept Baker's tripartite description of Old Regime political rhetoric as definitive, Sorbière's speech, a century before Turgot's attempted reforms, sought to set something very like Baker's "discourse of reason" in motion, not merely to define a new political outlook, but to define the means by which the troubled political world could be transformed.[5]

Revolution could not have been further from Sorbière's mind, however. Following the chaos of the religious wars in France, not to mention across the Rhine, and the civil and political disturbances of the Frondes, Sorbière, like many thinkers of his era, greatly desired the return of stability, of social *tranquility*, to use the term used in the time. Whether we understand the crises of the era as a series of events or as symptoms of one unitary event does not matter.[6] Sorbière and his colleagues articulated an intense and unsettling experience of social, spiritual, and political crisis. To them, France and even European society as a whole seemed on the verge of collapse, having descended into a murderous madness. The question was what, if anything, could be done—a problem posing something very like what Paul Hazard termed a "crisis of the European mind."[7] Sorbière believed he had the answer, however: social and political tranquility required a strong, even despotic, government buttressed by the authority of scientific reason. Agreeing with Thomas Hobbes, with whom he corresponded, Sorbière argued that a despotic monarchy was the best proof against chaos—and, furthermore, that the natural sciences would support the monarchy in creating a rational polity in which tranquility could reign.

Although we can discern its outlines as far back as the late sixteenth century, with the Stoic revival, a political discourse of scientific reason began to emerge with clarity in the 1660s among the savants of Paris, driven by thinkers and organizers such as Christiaan Huygens, Charles Perrault, and Sorbière. Sorbière's belief (or was it desperate hope?) that the sociopolitical order could be scientifically rationalized became a central feature of the Enlightenment philosophy later promoted in a far less modest manner by such philosophes as Voltaire and Condorcet (as I will detail later in this book). The close association of Sorbière's rationalist political discourse with notions of enlightened despotism, as we might call it, or *despotisme légal*, to use a contemporary term that became popular by the early eighteenth century, did not end until the disillusioning final decades of the eighteenth century, particularly after the fall of Turgot in 1776, when it became difficult to reconcile the goal of rationalization and stability with a failing monarchy.[8] Only then did surviving philosophes, especially the younger and even more radicalized second generation who had little of the intellectual modesty of their forebears, begin to abandon the principle of monarchy and promote a stronger, more centralized state formation—a technocratic republic administered by an educated elite that could sweep away what they had come to believe were the irrational and unjust vestiges of the *Ancien Régime*. Indeed, I think the hubris of philosophes such as Condorcet by the 1770s is not inconsequential to understanding their political radicalization.

Although historians have used various structural and poststructural theories to explain the origins of revolutionary ideas and attitudes, such explanations often seem to hinge on the unveiling of some true actor (the structure itself) operating beyond the consciousness of people whose lack of agency is disguised behind some false consciousness. In such explanations, the Revolution always seems to be an inevitability, like the dialectic between the state and the public sphere in Jürgen Habermas's once popular theory.[9] Sorbière's support for despotism, however, was not backward, or a false consciousness, or a betrayal of the true interests of the authentic public sphere as it existed in the Republic of Letters. Instead, Sorbière helped to initiate a critical strand of pro-despotic Enlightenment thought that furthered Neostoic ideals and was later upheld much more forcibly by the likes of Voltaire and Turgot, only collapsing with Condorcet. The Montmor Discourse, I think, provides a new perspective (more from the past than the future) on the geneology of the Enlightenment, viewing it as a response to the catastrophic (and least as they were so perceived) events of the sixteenth and early seventeenth centuries.

The desire for social and political order during the crises of the late sixteenth and early seventeenth centuries was a motivating force for proponents of a political discourse of scientific reason. Sorbière, in his spring 1663 speech, sought to prepare the informal Montmor Academy for formal status in the state because he thought the sciences offered the state practical assistance and the authority of reason. Praising Montmor for his support of the sciences during the dark years, Sorbière stated, "There is nothing more to wish for [Montmor's] glory and for the public utility, if he never engages in another enterprise so noble, than that it should become an enterprise of Sovereigns."[10] Elitist, pessimistic in general about human nature, and distrustful of the nobility, Sorbière saw a despotic regime advised by a ministry of experts and scientific institutions as the best hope for establishing a tranquil, disciplined, and orderly society. We can best understand Sorbière's political views as an elaboration of the ideals of the *politiques*, the religious moderates who, beginning in the late sixteenth century, desired to bring an end to war and anarchy through religious toleration and a strong monarchy with absolute sovereignty.[11]

The *politiques* of the sixteenth century found the philosophical expression of their political views in a revival of both ancient Greco-Roman Stoicism and skepticism.[12] By promoting above all the virtues of tranquility, moderation, and wisdom, Neostoicism offered a moral *via media* to those disinclined to take a dogmatic stance during the religious wars. The Stoic's apathy toward the external world also complemented the skeptic's suspension of judgment.[13] If absolutely certain knowledge of the external world is impossible, Neostoics argued, the wise should opt to suspend judgment and cultivate a Stoic apathy (*apatheia*) toward the unknowable and uncontrollable.

To *politiques*, the political significance of the Neostoic and skeptical positions lay in their denial of any religious justification for civil or religious strife, which they blamed on the immoderate passions that stemmed from errors of judgment. The Belgian refugee Justus Lipsius, in his hugely popular Neostoic work, *De constantia libri duo qui alloquium praecipue continent in publicis malis* (1584), argued that a philosopher achieves wisdom through reason, and thus learns virtue and tranquility. Wisdom keeps philosophers' souls free from immoderate passions by allowing them to distinguish accurately truth from falsity.[14] Bernard de Fontenelle, the greatest defender of the sciences in the late seventeenth and early eighteenth centuries (see chapter 2), would later argue that philosophers achieve wisdom (*la sagesse*) through a study of natural philosophy, thus implying that all philosophers must also be savants.[15] The wise, Lipsius suggested, obey the laws of the state scrupulously and do not get involved in conflicts that disrupt social or political order and thus prevent contemplative tranquility.[16]

The *politiques* found a French interpretation of Lipsius in the works of Guillaume du Vair. As bishop of Lisieux and later Keeper of the Seals under Louis XIII, du Vair lived a public life and adapted Neostoic themes to his own civic humanism. Like Lipsius, du Vair advised a purging of the violent passions that led to error.[17] When the state is well ordered, a Stoic philosopher's sphere of action is his or her own soul. In times of trouble, however, du Vair argued that the wise must act vigorously in the world to remedy the situation.[18] According to du Vair, the duty and responsibility of the virtuous and talented is to help lead society toward tranquility, because no one can find happiness in solitude if one's fellows are in misery.[19]

Lipsius, du Vair, and other skeptics and Neostoics, including Michel de Montaigne, provided philosophical authority to the *politique* position, but their intellectual influence paled next to the most widely read Neostoic author of the era: Pierre Charron. Charron's *De la sagesse* appeared in thirty-nine editions between its first publication in 1601 and 1672. Praised and critiqued, the work was placed on the Index in 1606, and yet it was called by Gabriel Naudé, the librarian of Richelieu and Mazarin, the greatest book ever written after the Bible.[20] Although derivative of other Neostoic works, including those of du Vair and Montaigne, Charron's work was seminal to the intellectual identity of savants such as Sorbière.

Like du Vair, and even his friend Montaigne, Charron was a *politique*. His turn to Neostoicism, like that of other Neostoic thinkers, marked an overriding desire for political stability and peace—for tranquility. An *avocat* of the Parlement of Paris at the end of the sixteenth century, Charron moved in the circle of the *politique* leader Michel de l'Hôpital. In anger and indignation, Charron briefly sided with the Catholic League after the assassination of the Duc de Guise in 1588. Upon the assassination of Henry III, however, Charron returned to the cause of the *politiques* and remained thenceforth faithful to Henry IV, to whom he dedicated his first published work, *Les trois vérités* (1593). Charron later explained his temporary defection as the result of violent emotions that clouded his reason.[21] *De la sagesse*, he explained, resulted in part from Charron's intense desire to seek the causes of his actions. He concluded that uncontrolled passion had been the culprit—false judgment, the result of prejudice and opinion, not true knowledge, had led to a passionate response.[22] The goal of the wise, Charron argued, should therefore be the supreme virtue of self-control, which regulates the passions by reason and will and is achieved through contemplation and knowledge. The suspension of judgment, as the skeptics advise, is wise because it preserves the mind from prejudice. Virtue and morality come not from adherence to dogmatic and unexamined religious doctrine but from wisdom, or *la sagesse*—one likely reason *De la sagesse* was put on the Index.[23]

The primary duty of the wise, according to Charron, is to control their conduct in the world, which implies both a duty to the internal self—the maintenance of self-control—and a civic duty similar to what du Vair suggests. "To flee and hide oneself while having the means to profit others and aid the public," Charron wrote, "is to be a deserter, to bury one's talent, to hide one's light."[24] Like du Vair, Charron asserted that the virtuous cannot maintain their tranquility and liberty alone while the rest of the world is in chaos. The internal cannot be radically separated from the external.

Charron did not believe that those who can achieve virtue through the ethic of self-control are like the mass of humanity, however. They are *esprits supérieurs*. The majority are either *les esprits faibles* or *les esprits médiocres*.[25] The weak-minded can be ignored; they are born only to obey. Mediocre minds are active and dogmatic, both ruling and troubling the world. Superior minds prefer to stand aloof from the relentless bang and clatter *les esprits médiocres* cause, but they must get involved when the state is in need. They observe the laws and customs of the land not from habit but from reason, and even perhaps with a hint of cynicism. If order, tranquility, and stability are to be maintained, a powerful central authority must enforce the obedience of the masses. The philosopher easily sees through the illusions of the king's assumptions of divine right and absolute power but supports them because they are the best hope for establishing the authority of the monarchy and thus ending civil strife.[26]

The generation most influenced by Charron, many of whom earned the epithet *libertins* or *libertins érudits* (learned freethinkers), formed the nascent scientific community in Paris, and from them a rationalist moral-political discourse emerged. Sorbière belonged to the second generation of learned freethinkers who took their ideals as a matter of course and for whom self-conscious religious and epistemological skepticism was a way of life.[27] The first generation, writing mostly in the early seventeenth century, included Gabriel Naudé (1600–1653), François de la Mothe le Vayer (1588–1672), Pierre Gassendi (1592–1655), and such disparate personalities as Father Marin Mersenne (1588–1648) and the poet Théophile de Viau (1590–1626).[28] As disciples of Charron, the learned freethinkers thought of themselves as *esprits forts*—echoing Charron's term *esprits supérieurs*—intellectual elites surrounded by weak and mediocre minds. They were skeptical of theological and epistemological dogmatism and privately cynical of authority, but publicly they evinced orthodoxy and loyalty.[29] They had no desire to disrupt a tenuous social and political stability by inciting in the masses anything other than Catholic orthodoxy and unquestioning loyalty to the monarchy. Most held offices of some sort from Richelieu or Mazarin and believed it was the job of the state, according to its enlightened self-interest, to ensure peace, harmony, and tolerance.[30] The *libertins érudits* were, at least in spirit, *politiques*.

From the ranks of the *libertins érudits* in the early seventeenth century emerged a moral discourse that upheld what we might term a Baconian ideal of the sciences as pragmatic, rational, polite, and civilizing.[31] Some of the learned freethinkers, such as La Mothe le Vayer, adopted a nihilistic Pyrrhonism in response to the loss of religious certainty, but others, such as Gassendi, found their attitude unnecessarily pessimistic in light of recent scientific achievements such as those of Galileo. In the first two books of his *Exercitationes paradoxicae adversus Aristoteleos* (1624), Gassendi upheld skeptical arguments and insisted that a science of necessary causes revealing certain but nonevident truths about some supposed real world existing independent of our sense perceptions cannot exist.[32] Certainty is impossible and must be given up in favor of probability, because the ultimate nature of things cannot be perceived, although "it can be stated also that there are many things that can be known . . . only by an experimental science or one following appearances."[33] In short, a science must avoid metaphysics, and so it must also avoid the dogmatism that results from false certainties. Gassendi continued to promote his epistemological *via media* in the *Syntagma philosophicum*, still uncompleted at his death in 1655, and thus he contributed a major development to the idea of the sciences as morally enlightening: reasoned skepticism helps constrain the unruly passions through devaluing certainty.

This moral discourse of the sciences obviously emerged from political and social concerns. Gassendi's close friend, Mersenne, for example, defended the sciences as a source of stability in a chaotic world. In *La Vérité des Sciences contre les Sceptiques ou Pyrrhoniens* (1625), and in other works, Mersenne argued that a basic skepticism must be accepted, but he agreed with Gassendi that useful, positive, and reliable knowledge of the experiential world could be attained.[34] Mersenne also appears to have believed that pragmatic, nonsystematic, and nondogmatic natural philosophy could best foster a sense of unity and stability among the intellectual elites of Europe, thus providing a model of civil society for the political elites to follow.[35] Savants must therefore work together to discover useful knowledge and not fight over inflexible beliefs or the doctrines of ancient authorities. Through such collaboration, the intellectual community could take up the social and political duty du Vair and Charron prescribed. Mersenne himself took this duty seriously: he carried out a vast correspondence with scholars throughout Europe, sponsoring cooperation and pragmatism. He and other great correspondents, such as Nicolas-Claude Fabri de Peiresc and Henry Oldenburg, created the post-Renaissance Republic of Letters in the early seventeenth century.

Mersenne, from the 1630s until his death in 1648, also organized a circle of savants who met to discuss matters of natural philosophy. The group, which at times numbered over sixty, including Descartes, Gassendi, Gilles Personne de Roberval, Christiaan Huygens, Pierre de Carcavi, and the Pascals (Étienne and Blaise), was cosmopolitan, transcending religious, social, and political differences. After Mersenne's death, the survivors of the group eventually made their way, virtually en masse, to the house of Montmor.[36]

And so, by a circuitous route, we find ourselves back to Sorbière's discourse. Soon after the 1648 death of Mersenne, savants such as Ismaël Boulliau, Blaise Pascal, Gilles de Roberval, Girard Desargues, Pierre de Carcavi, and Sorbière began congregating irregularly for discussions and dinners at the residence of Montmor, a wealthy officeholder and founding member of the Académie Française. By the early 1650s, Montmor was the leading supporter of the scientific community in Paris, and from about 1657 on, Montmor scheduled regular weekly meetings for scholars to discuss scientific matters.[37] He even provided lodging for Cartesianism's aging philosophical opponent, Pierre Gassendi, which made the Montmor house especially attractive to savants.[38] After Gassendi's death in 1655, Gassendi was buried in the Chapelle de Montmor, and Montmor himself led the organization and publication of Gassendi's *Oeuvres* in collaboration with other savants, including Sorbière.[39] By placing the interests of the Republic of Letters above petty internecine disputes, Montmor won the respect of savants, who longed for a society in which stability and reasonableness replaced anarchy and dogmatic inflexibility.

The disharmonies of society, of human nature in general, however, were replicated among the supposed *esprits forts* of the Montmor Academy. By 1663, disputes over philosophical systems, selfish concerns with personal glory at the expense of scholarly exchange, and personal animosities threatened to destroy the Montmor Academy despite Montmor's best efforts to discipline the institution effectively with rules and regulations—despite his efforts to pose, in fact, as a petty, enlightened despot in a tiny realm of savants. He soon grew bored with the whole affair and had other problems to attend to when several savants, led by Sorbière, made a last-ditch effort in April 1663 to renew the academy's spirit of cooperation.[40]

Sorbière, in his speech on April 3, expressed his annoyance at the dissonances of the Montmor Academy, which he took personally—in 1657, upon Montmor's request, Sorbière and his colleague, Abraham du Prat (deceased in 1663), had drawn up the academy regulations.[41] Some members thought themselves above etiquette and openly flaunted the rules, Sorbière complained, and ruined the academy assemblies for the rest by willfully disrupting meetings. Others, whom Sorbière termed *qui non scire cupiunt ut sciant, sed ut sciantur*—we might say "windbags"—wasted everyone's time with useless, frivolous discourses. Such speechifying broke the first article of the Montmor Academy regulations, upon which the pragmatic, anti-Aristotelian spirit of the academy depended: "That the purpose of the Conferences will not at all be the vain exercise of the mind on useless subtleties, but will always focus on the clearest knowledge of the works of God, and the improvement of the conveniences of life in the Arts and Sciences which serve better to establish them." These windbags, Sorbière complained, associating them with the Aristotelians of the universities, "have come here only to kill time and to acquire esteem." They made no significant contributions to the advancement of knowledge.[42]

Sorbière's attack on the speechmakers did not imply that he supported the elimination of all learned discourses, as some exasperated experimentalists had advocated. He himself preferred the experimentalist method, but an elimination of all discourses would, he argued, break "our primary resolutions, which are to carry out a learned and judicious mix of experiments and reasoned discourses." Moreover, an orderly and harmonious society, best suited to foster progress and peace, would exclude "no method, provided that it does not contravene order."[43] There was no virtue in adopting a systematic or methodological dogmatism.[44] A determined and self-conscious skepticism, Sorbière believed, remained critical to the maintenance of order.

Besides the windbags, Sorbière also complained about those savants who willfully behaved with incivility. Noting that certain savants, treated with "too much indulgence," had disrupted the academy soon after its establishment, Sorbière insisted on the necessity of depriving the academy of their enlightenment, rather than allowing them to silence others. He was likely referring to the brilliant but vain mathematician, Gilles Personne de Roberval, who, in 1658, had argued rudely with Montmor and was subsequently not invited to any further assemblies.[45]

For Sorbière, the maintenance of order was an issue of sociability and politeness, but not for their own sake—the unfettered communication of practical knowledge depended on the control of irrational and disorderly passions. "How many times," complained Sorbière, had the Montmor Academy witnessed "two men who are with us no longer, two minds as penetrating and judicious as there ever were to my knowledge, two disciples of Mr. Hobbes and two copies of Bacon, the wise M. du Prat, and the agreeable M. du Bosc, leave without having said a single word because two others had amused fruitlessly this company for three solid hours, as if they were the only ones with something good to say?"[46]

The academy was a microcosm of French society; it required authority, regulation, and discipline to function properly. Effective and practical—that is, enlightened—rules ensured tranquility, enabled communication, and facilitated the progress of knowledge through justice. Unlike society at large, however, the academy, as a community of *esprits forts*, should be able to regulate itself—but it was failing. We should be careful not to diminish the social and political awareness of these savants by reducing the informal scientific associations to sites of patronage and aristocratic politesse. The great nobility had disgraced itself in the recent Frondes, and some savants held it to blame for the turmoil of the past century and a half. Sorbière sought to advance the dream of Mersenne and create in the scientific community a new social paradigm that favored tolerant skepticism rather than dogmatism and rational, intellectual association rather than self-interested designs for glory and honor. His frustration and despair grew steadily as this dream teetered toward collapse.

Sorbière had a larger vision of the role the sciences, and savants, should play in the state. To satisfy this vision, he hoped to make the academy more attractive to the monarchy and to convince unconvinced savants that they needed the formal recognition of the state. The Montmor Academy, Sorbière insisted, showed that private scientific associations had gone as far as possible in promoting a study of the sciences. In order to complete their work, savants needed such funding, facilities, laboratories, machines, and manpower as even wealthy patrons like Montmor could not provide: "In truth, Messieurs," said Sorbière, "only Kings, and rich Sovereigns, or some wise and wealthy Republics could undertake to outfit a Science Academy engaged in continuous experimentation," and until they did, "our Mechanics will remain imperfect as they are, our Medicine will be blind, and our Sciences will teach us with certainty only that there is an infinity of things that we know nothing about."[47]

We have long known that the French monarchy made a determined effort to bring intellectual activity in France under its control in the post-Frondes era.[48] A truly absolute monarchy desired not only the centralization of military, legal, and political power in France but also the control of intellectual and religious forces.[49] The establishment of royal academies in various fields of expertise, including the natural sciences, to channel intellectual and artistic production to the service of the state was part of this program.[50] Savants did not necessarily perceive such centralizing practices as oppressive or authoritarian. From 1648 to 1653, during the turmoil of the Frondes, many savants turned to the monarchy, just as their Neostoic predecessors had during the religious wars, as the only hope for order. Sorbière, who, from his home in Leiden, followed with horror news of the depredations of the Frondes, became an enthusiastic and vocal proponent of an absolute, even despotic monarchy—sentiments that, as an admiring follower of Thomas Hobbes's philosophy of centralized state power, he had expressed earlier in the 1640s, in the last glory days of Richelieu and Louis XIII.[51] In 1649, with the Frondes under way, Sorbière translated into French Hobbes's *Elementa philosophica de cive*, followed in 1653 by the *Corpus politicum*.[52] In the 1650s, Sorbière's outrage inspired him to compose tirades promoting despotism. In 1656, for example, he wrote two essays: the first, *Discours Sceptique à Philotime, pour montrer que Paris et les Français ne sont pas tout à fait exempts de barbarie*, ranted about the terrible state of affairs in Paris and France; the second, *Discours Sceptique en faveur des bêtes et du gouvernement despotique*, insisted that men live more happily under a despotic government than under one less absolute.[53] Far from seeing despotic regimes as repressive, Sorbière argued that they protect the people, who "are all [then] immediately under the protection of their Sovereign," from mistreatment at the hands of an autonomous nobility. An insufficiently despotic government, Sorbière wrote, allows "a governor of a region or province" to disobey "sometimes with impunity the orders of the sovereign," while "the least folly of the people is severely punished, as if there is not more to fear from the disobedience of the great than from the little people."[54]

By the mid-1650s, if not earlier, Sorbière had come to believe in the necessity of a rational but despotic government if France were ever to achieve political stability and to promote justice. Moreover, he was convinced that the arts and sciences, by promoting civilization and reason, both supported and required a strong state—and he strove to direct the energies of the Montmor Academy toward his desired ends. His idealism is evident in a 1658 letter to Thomas Hobbes, who, intrigued by reports of a new science academy with formalized rules of etiquette, the first in France, had requested a copy of the regulations Sorbière and du Prat had only recently devised. Eager to obtain Hobbes's approval, Sorbière sent a transcription of the nine articles and waxed poetic about a future when the arts and sciences promoted peace and harmony on Earth: "The age of iron will not last forever. Peace will return in its time to the Earth, the Muses will not be eternally exiled, the Arts will be reborn, the Sciences will once again take their place, and a more gentle influence of the Stars than that which today produces only soldiers and Captains, will produce once again the Gilberts, Bacons, Harveys, Fra Paolos, Galileos, Mersennes, Descartes, and Gassendis."

Institutions such as the Montmor Academy, Sorbière believed, would prove "the glory of our age," enabling "these Illustrious persons to teach each in their turn that, despite the barbarism in which we have lived, there is found a great number of honest men who have not lacked love of Philosophical studies." Sorbière hoped he would be able to shape the Montmor Academy to realize his utopian vision: "God grant that I be a Prophet with my desire that we imitate the Modesty of Monsieur Gassendi, that his sweet and tranquil spirit reign in our Assembly and that this new Academy fail not the hopes of these regulations that I send to you."[55]

By the end of the 1650s, Sorbière had begun to articulate and act on the idea that the world should conform to the dictates of scientific reason—indeed, that the natural sciences must become the organizing principle of the polity. And to reiterate, Sorbière's political philosophy, although consistent with his education, experience, and ideals, was not at all democratic, egalitarian, or revolutionary. It was pro-despotic. And in 1658 it seemed stillborn. Sorbière was not yet the prophet he hoped to be in his letter to Hobbes, and problems of organization, regulation, and philosophy plagued the academy nearly from its beginning. Later in that year the infamous Roberval incident cast a pall over the meetings. But Sorbière doggedly continued his efforts to keep the academy focused. In a June 1658 speech, he admonished the academy savants to embody the practical spirit of Gassendi: "I do not want us to dispute eternally on what fire is. Rather I wish that we would report on new inventions, or on the practices of diverse lands, for warming oneself better."

An easier and more convenient life, and hence a happier one, Sorbière insisted, contributed to peace and tranquility, and since these were the goals of the state, the sciences should soon attract the attention of the monarch and his ministers.

> We have seen in this Assembly the first men of the Robe, the Cordons Bleus, Dukes and Peers, and Grand Prelates. Our scientific researches find themselves honored by their presence. Their example could well advance our discoveries and favor the commodities of life if Sovereigns also stood one day before the land, and claimed afterwards that they had extended and assured the felicity of their People—to which, without doubt, the natural sciences, and the Arts that depend on them, could contribute much—certainly if we could by their means make our health better, lives longer, and happiness more universal. Such would be a rather magnificent plan.[56]

Sorbière was not content merely to harangue the savants of the Montmor Academy. He also cultivated friendship and patronage in high places, particularly with Cardinal Mazarin. Since converting from Protestantism to Catholicism in 1654 (following the example of the famous Neostoic, Justus Lipsius—not to mention Henry IV!), Sorbière had maintained an irregular correspondence with Mazarin, and from 1657 onward he sought to focus Mazarin's mind on the utility of the natural sciences to the state.[57] In 1659, Sorbière collected his various discourses and letters, many to Mazarin, on philosophy and the sciences, and he published them in two volumes that were dedicated to the cardinal.[58] The dedication, which strongly echoed Sorbière's optimistic 1658 letter to Hobbes concerning the imminent conclusion of the "age of iron," contained a not-so-subtle injunction for Mazarin to support the natural sciences (and the savants who practiced them), which would then support him and his glory:

> All the Sciences and Arts rejoice to see you return to your first inclinations: And I have no doubt that, after the great discoveries that the Galileos, Gilberts, Bacons, Harveys, Gassendis, Hobbes, Descartes, and several other excellent men have made, that if you favor them, and if you undertake to push them vigorously much will be accomplished before the end of the century. This would not make, in my opinion, a small mark in posterity for the glory of Your Eminence. But the Sciences have no need that I speak for them, and you have made common cause with them. The Friendship that is between you and them is too venerable, and your interests are confounded with theirs, to allow me any pretext of representing them to you.[59]

Curiously, Sorbière's ideas about the utility of the natural sciences to the state were growing more complex than a cursory reading of his dedication might reveal. In Sorbière's view, the sciences were not only useful for the innovation of new technologies, but they could also provide rational authority to state power. When Sorbière suggested that Mazarin's interests "are confounded" with the sciences, he implied that politics can be made scientifically rational and that such rationality supports centralized authority.

Mazarin probably knew exactly what Sorbière meant. In an earlier letter to Mazarin, Sorbière had explicitly argued that politics could be studied scientifically, using similar a priori proofs one finds in mathematical sciences such as geometry. Sorbière thus contradicted the common belief that politics, a creation of human cultures, could never be explained or predicted by definitive scientific proofs, which ought to be based on universal principles. On the contrary, Sorbière argued, the science underlying politics is ethics—the science of what is just and unjust—and the proofs of ethics are, in fact, known a priori. One cannot induce general ethical propositions from the particular behaviors of humans in a state of nature because for such humans justice and injustice have no meaning. Their social activity is no different from that of any social animal. Justice and injustice, rather, are deduced from their causes, which are the laws and pacts people have made in accordance with their reason. One cannot understand justice by observing human behavior, but rather by consulting one's reason and deducing the manner in which people *ought* to behave, and then making laws to regulate that behavior. Similar to Hobbes, Sorbière argued that justice, law, and right do not emerge naturally from society; a rational authority must impose them from above. Mazarin was likely to favor such an argument linking scientific rationalization with royal authority.[60]

Sorbière's obsequiousness served him little, however. Mazarin died in March 1661, and Sorbière lost his privileged connection within the government. To make matters worse, failure continued to dog his efforts. He was less familiar with Mazarin's protégé, Jean-Baptiste Colbert, who founded the Académie royale des sciences in 1666. Several of Sorbière's colleagues—including Christiaan Huygens, who was by then one of the most respected savants in Europe—also advocated royal patronage of a science academy and were closer to Colbert. They soon stole the limelight, especially after the appearance of the Project de la compagnie des sciences et des arts, a proposal for a broadly conceived academy of human knowledge. Although usually associated with Huygens, the project appeared only a few days after Sorbière's 1663 speech to the Montmor Academy, and there is no reason to assume that Sorbière was ignorant of it.[61] Certainly the project reached Colbert, probably through Huygens, and became the basis for later formal proposals for a royal academy of the sciences.[62] In the early 1660s, the clamor for an official science academy was reaching a crescendo, and Sorbière was only one of the more vocal advocates.

A few weeks after his speech of April 3 to the Montmor Academy, Sorbière sent a copy with a cover letter to Colbert himself. "Permit me," began Sorbière, "to treat with you as I did with His Eminence [Mazarin]." Sorbière noted that he desired nothing for himself but to pass the rest of his life in "the business of the Sciences," and he sought to impress upon Colbert, as he had Mazarin, the importance of supporting scientific endeavors, "as you will see from my harangue of 3 April, which could lead to something important for the public if it is considered by those who work for the ornamentation of France."[63] Linking the utility of the sciences with the health of the state, Sorbière asserted that Colbert could not heal the state of its maladies, nor bring it to its full stature, unless he helped the arts and sciences to flourish.

In the *Sorberiana* (1694), the collected sayings of Sorbière, his admirers sought to remind the public of his role in fostering the *goût des sciences* that swept France at the end of the seventeenth century.[64] Since the foundation of the Académie royale des sciences twenty-eight years earlier, a craze for things scientific had taken hold of polite culture, and scientific experiments and writings were the rage in the salons of Paris.[65] The sciences were not merely to entertain the wealthy, however. As the rising voice of the scientific community, Fontenelle had implied eight years earlier in his popular *Entretiens sur la pluralité des mondes* (1686) that the sciences could fundamentally alter human consciousness. In suggesting that Earth was only a small part of a vast universe filled with many inhabited planets, Fontenelle implicitly argued for a rational worldview that would tolerate differences and abhor religious arrogance and dogmatism.

The sciences offered human society new ways to construct morality and politics, which the compilers of the *Sorberiana* believed had been Sorbière's message as well—as *esprits forts*, the duty of savants, per the Neostoic tradition, was to promote tranquility and peace, which meant supporting a strong central authority, the monarchy of Louis XIV. Such support, while pragmatic, was also ideological. In the *Sorberiana* Sorbière begins to sound like the fiercely rationalist philosophes of the late eighteenth century: "The Mathematical Sciences are the true Metaphysics and Moral Science is the true Theology," he is quoted as saying. "They both arise from the Physical Sciences, and civil society is their invention because our Arts and our Customs come from them, and they draw out of us all those things that distinguish us from other animals . . . I avow to you that Moral Science and Mathematics come together, and that Physics is an Ocean to which all our knowledge will return."[66]

Although Sorbière's vision places him at the beginnings of Enlightenment thought, far from standing in glorious opposition to despotism, he is uncompromisingly in favor of it. The philosophes' eventual role in the development of a revolutionary consciousness, then, must be understood at least as much from an ideological as from a social-structural perspective—one, moreover, that reads historical change historically rather than proleptically. Sorbière's Montmor Discourse was a part of the post-Frondes anxieties that saturated French intellectual culture. Through the Discourse and his other actions, Sorbière sought to resolve the disjuncture between the chaos of the sociopolitical world and the supposed tranquility of the scientific world by reforming the monarchical state to conform to the dictates of scientific reason and by putting savants at the center of reform. By the mid-1770s, the catastrophic failure of this reform project, as attempted by later philosophes, may have helped to spark revolutionary sentiments, but even in the 1660s, the idea of the morally enlightening sciences was on the verge of collapse. Sorbière himself was hounded by failure and despair, and he seems even in his own estimation to have achieved little or nothing, despite constant effort.

In summer 1663, soon after his letter to Colbert, Sorbière departed on a voyage to England, where he visited Hobbes and attended meetings at the famous Royal Society—thus, perhaps, initiating the Anglophilia of the philosophes made most famous by Voltaire sixty years later.[67] Upon his return to France, Sorbière wrote an account of his travels, which appeared in 1664 under the title *Relation d'un voyage en angleterre*. Predictably, the account focused upon the Royal Society in a pointed comparison to what Sorbière considered the failings of the Montmor Academy. Of Royal Society fellows, Sorbière wrote approvingly, "They report in as few words as they find appropriate to explain the experiments that the Secretary has proposed. No one hastens to speak, nor takes it upon himself to be longwinded and say everything he knows. They never interrupt whoever is speaking." The Royal Society's motto, *Nullius in verba*, expressed perfectly Sorbière's skepticism of philosophical systems. Despite arguments over theories, hypotheses, and principles, Sorbière insisted, all members of the Society behaved toward each other with such moderation that their work was never impeded. The fellows "know well that they are coming together for the same end, even if by different routes, since they all work to explain the same phenomena."[68]

Sorbière foresaw the result of collaborative, government-supported scientific work as "an infinity of useful inventions" for everyone, leading to general peace and prosperity as life improved. A monarch who supported such work, Sorbière added, would receive glory far beyond those who seek it in "ill-considered wars" and would be remembered forever. The glory of the warmongers, he continued, "is almost completely forgotten, and no moderate person will read except with horror of the desolations they have caused . . . Man is not born to wage war against his fellows, but to enjoy the fruits of peace."[69]

By adding his voice to the call for a Royal Academy of Sciences, Sorbière was instrumental, even if indirectly, in the final push to found one. By 1664, Perrault and Huygens had Colbert's ear regarding a general learned academy. After the spectacular comet of 1664, astronomers such as Adrien Auzout were able to make a compelling case for the creation of a royal observatory. Clearly in a mood to listen, Colbert lent his Parisian residence for observations of the eclipse of 1666, and by that summer he was selecting the first members of the official Académie royale des sciences.[70]

Sorbière, however, enjoyed no part in these grand events. In June 1664, despite his efforts, the Montmor Academy collapsed. Although Huygens hoped that a royal academy would soon arise from its ashes, the end of the Montmor Academy marginalized Sorbière in the Parisian community of savants.[71] Adding to his personal discouragement, his scientific colleagues in England received his *Relation* poorly, and his dream of savants working together for the common good took yet another blow. An apparently indiscreet remark about the lord chancellor brought an open complaint from the English government, and in an effort to retain good relations across the channel, the French government issued an *arrêt de Conseil d'Etat* on July 9 ordering *Relation* suppressed and its author banished from the capital.[72] Although granted royal clemency in October, Sorbière found he had earned the nationalist rancor of his colleagues in the Royal Society. In 1665, Thomas Sprat, historian of the Society, issued a violent, rambling, and incoherent reply to Sorbière's *Relation* titled *Observations on Monsieur de Sorbier's Voyage into England*, and he sought to have Sorbière's name stricken from the Society rolls in 1666.[73]

Discouraged and ill, Sorbière committed suicide on April 9, 1670, by overdosing on laudanum. In the end, Sorbière's despair destroyed him, but his reaction to the condition of post-Frondes France led him to articulate a new relationship between science and polity—one that would have influential and interesting ramifications in later decades and that would propel savants to the center of the political maelstrom.

NOTES

1. See David Sturdy, *Science and Social Status: The Members of the Académie des Sciences, 1666–1750*, 3–25; Guillaume Bigourdan, "Les premières réunions savantes de Paris au XVIIe siècle—L'Académie de Montmor," 159–62, 216–18; Bigourdan, *Les Premières Sociétés Savantes de Paris et les Origines de l'Académie des Sciences*; and Harcourt Brown, *Scientific Organizations in Seventeenth Century France (1620–1680)*, 64–90.

2. Charles Coulston Gillispie, *Science and Polity in France at the End of the Old Regime*.

3. Baker, "Science and Politics at the End of the Old Regime," in *Inventing the French Revolution*, 153–60. See also C. B. A. Behrens, *Society, Government and the Enlightenment: The Experiences of Eighteenth-Century France and Prussia*, and Julian Swann, "Politics and the State in Eighteenth-Century France," in T. C. W. Blanning, ed., *The Eighteenth Century: Europe 1688–1815*, 11–51.

4. In Baker, *Inventing the French Revolution*, 12–27.

5. Challenging Baker's formulation, Robert Darnton makes a compelling alternative suggestion in *The Forbidden Best-Sellers of Pre-Revolutionary France*, 179.

6. See Jonathan Dewald, "Crisis, Chronology, and the Shape of European Social History," 1051–52.

7. Paul Hazard, *La crise de la conscience europèene (1680–1715)*.

8. *Legal despotism* was a physiocratic term, also cited approvingly by Turgot in 1767. See Turgot to Du Pont de Nemours, November 22, 1767, in *Oeuvres de Turgot*, 2: 677.

9. See Jürgen Habermas, *The Structural Transformation of the Public Sphere: An Inquiry into a Category of Bourgeois Society*; Dena Goodman, *The Republic of Letters: A Cultural History of the French Enlightenment*; and Daniel Gordon, *Citizens without Sovereignty: Equality and Sociability in French Thought, 1670–1789*.

10. The discourse is preserved at the Bibliothèque Nationale, Manuscrits, *Cinq Cents de Colbert*, n° 485. It is printed in full in G. Bigourdan, "Les premières réunions savantes de Paris au XVIIe siècle—L'Académie de Montmor," 159–62, 216–18. Unless otherwise noted, all translations are mine.

11. On the development of absolutist thought among the first *politiques*, see N. O. Keohane, *Philosophy and the State in France: The Renaissance to the Enlightenment*, chapter 2, especially sections regarding Chancellor Michel de l'Hôpital and Jean Bodin.

12. See Gerhard Oestreich, *Neostoicism and the Early Modern State*, and R. H. Popkin, *The History of Scepticism from Erasmus to Spinoza.* As Popkin argues, the crisis of the Reformation was not only religious and political but also intellectual. Once the Reformation had challenged the traditional criterion of religious knowledge—and ultimately all knowledge—certainty had to be given up. The revival of skepticism and Stoicism was part of the rational response to this intellectual crisis.

13. See Anthony Levi's discussion of this in *French Moralists: The Theory of the Passions, 1585–1649*, 11.

14. Levi, *French Moralists*, 68. On this general subject, see also Eugene Rice, *The Renaissance Idea of Wisdom*, and George Huppert, *The Style of Paris: Renaissance Origins of the French Enlightenment.*

15. See also G. M. Adkins, "When Ideas Matter: The Moral Philosophy of Fontenelle," 433–52.

16. Lipsius also outlines a Neostoic political philosophy in the *Politicorum, sive civilis doctrinae libri sex* (1588), echoing the *politique* argument that *raison d'état* should guide a ruler in his effort to establish social harmony and political tranquility. See Levi, *French Moralists*, 63–73.

17. See du Vair, *La Sainte Philosophie* (1584), and *La Philosophie morale des stoïques* (1585).

18. Because of his focus on action and honor, du Vair had great influence on the heroic ethic in the plays of Pierre Corneille. See Paul Bénichou, *Morales du Grand Siècle.*

19. Du Vair, "Exhortation à la vie civile," in *Les Oeuvres du Sr. du Vair au Parlement de Provence*, 800–806. See also Levi, *French Moralists*, 74–95.

20. Levi, *French Moralists*, 95–96.

21. Levi, *French Moralists*, 97.

22. Charron, *De la sagesse*, book 1, chapter 18, §§ 1–3 in *Toutes les Oeuvres de Pierre Charron*, 65–67.

23. Charron, *De la sagesse*, book 2, chapter 2, §§ 1–2, 11–12.

24. Charron, *De la sagesse*, book 1, chapter 54, § 4, 188. Cited in Keohane, *Philosophy and the State*, 136.

25. Charron, *De la sagesse*, book 1, chapter 13, 153–55.

26. Charron, *De la sagesse*, book 2, chapter 8, 81–90.

27. For example, Sorbière applied the term *sceptique* to various discourses more to indicate his philosophical position than to talk specifically about skepticism. See the two mentioned below, as well as the *Discours sceptique sur le passage du chyle, & le mouvement du coeur . . .* He also prepared a *Lettre sur le Pyrrhonisme* in 1660 and began (but never finished) a translation of the ancient skeptic, Sextus Empiricus, in 1656.

28. Seventeenth-century French *libertinage* encompassed a range of moral and intellectual ideas and thus cannot be described as a single movement, although they shared enough characteristics that those of less critical orthodoxy sometimes feared they were engaged in a conspiracy to undermine traditional faith. See René Pintard, *Le Libertinage Erudit dans la Première Moitié du XVIIe Siècle*, 2 vols., and Popkin, *The History of Scepticism*, chapter 5.

29. Keohane discusses at length the "two moralities" of philosophers, public and private, in the sixteenth and seventeenth centuries in *Philosophy and the State*, especially chapter 4.

30. For example, Gabriel Naudé was librarian to Richelieu and Mazarin and secretary to Cardinal Bagni; Gui Patin was rector of the medical school of the Sorbonne; Leonard de Marandé was Richelieu's secretary; La Mothe le Vayer was tutor to the royal family; and Gassendi was a professor of mathematics at the Collège Royal. Sorbière later held pensions from Mazarin. Popkin, *History of Scepticism*, 87.

31. Recent historiography has challenged the influence of Bacon in the seventeenth century but, in the context of France, savants did conjure Bacon's name when highlighting the social utility of the sciences. See Peter Dear, *Discipline and Experience: The Mathematical Way in the Scientific Revolution*, 2–9, 235; and Steven Shapin, *A Social History of Truth*, chapter 7. See also Popkin, *History of Scepticism*, chapters 5 and 7; Wendy Perkins, "The Uses of Science: The Montmor Academy, Samuel Sorbière, and Francis Bacon," 155–62; and Michèle le Doeuff, "Bacon chez les grands au siècle de Louis XIII," 155–78.

32. Pierre Gassendi, "Dissertations in the Form of Paradoxes Against the Aristotelians, Sixth Dissertation (1624), art. 1," excerpted in *The Philosophy of the Sixteenth and Seventeenth Centuries*, ed. Richard Popkin, 111.

33. Book II, sixth dissertation, article 1, article 6, excerpted in Popkin, *The Philosophy of the Sixteenth and Seventeenth Centuries*, 110–13.

34. On Mersenne's theory of science, see Popkin, *The History of Scepticism*, chapter 7, and R. Lenoble, *Mersenne ou la Naissance du Mécanisme*. Also Peter Dear, *Mersenne and the Learning of the Schools*.

35. On this theme, see Sturdy, *Science and Social Status*, 14–16.

36. See Brown, *Scientific Organizations*, chapters 2 and 3, and Robert Mandrou, *From Humanism to Science, 1480–1700*. See also Peter Miller, *Peiresc's Europe: Learning and Virtue in the Seventeenth Century*.

37. On Montmor's life and patronage of the sciences, see Suzanne Delorme, "Un Cartésian Ami de Gassendi: Henri-Louis Habert de Montmor," 20–31.

38. The relevant historiography on the controversy between Gassendi and Descartes, and later on Gassendism and Cartesianism, is huge. A good starting point is Thomas Lennon, *The Battle of the Gods and Giants: The Legacies of Descartes and Gassendi, 1655–1715*.

39. Sorbière wrote the preface and added his *De vita et moribus Petri Gassendi* to the Lyon edition of Gassendi's *Syntagma philosophiae epicuri* (Lyon, 1658).

40. Montmor experienced various difficulties in the 1660s, which ended disastrously in 1669 when he was forced to sell his charge of Maître des requêtes to pay the debts of his eldest living son. Delorme, "Un Cartesian Ami de Gassendi," 68.

41. Sorbière copied the articles in a letter to Hobbes on February 1, 1658. See Sorbière, *Lettres et Discours de M. Sorbière sur diverses Matieres Curieuses*, letter LXXIX; Huygens, *Oeuvres Complètes*, N. 455a, 4: 513–16. English translation, Brown, *Scientific Organizations*, 75–76.

42. Bigourdan, "Les premières réunions savantes de Paris au XVIIe siècle—L'Académie de Montmor," 162.

43. Bigourdan, "Les premières réunions savantes de Paris au XVIIe siècle—L'Académie de Montmor," 162, 216.

44. Roger Hahn shows that there was no methodological solidarity within the Academy itself—only an agreement to uphold "philosophical pluralism." See R. Hahn, *The Anatomy of a Scientific Institution: The Paris Academy of Sciences, 1666–1803*, 112–15.

45. The incident is famous mostly because the gregarious astronomer Ismaël Boulliau afterward described the clash in a letter to Christiaan Huygens. *Oeuvres de Huygens*, n. 553, 2:286–87: "As for M. de Roberval, he has done a very stupid thing in the house of M. de Montmor who is as you know a man of honor and position; he was so uncivil as to say to him in his own house, having taken offence at an opinion of M. des Cartes which M. de Montmor approved, that he had more wit than he, and that he was less only in worldly goods and the office of Maître des Requêtes, and that if he were Maître des Requêtes he would be worth a hundred times more. Monsieur de Montmor, who is very circumspect, said to him that he could and should behave more civilly than to quarrel with him and treat him with contempt in his own house. The whole company found the boorishness and pedantry of M. de Roberval very strange." Translation from Brown, *Scientific Organizations*, 87.

46. Bigourdan, "Les premières réunions savantes de Paris au XVIIe siècle—L'Académie de Montmor," 161.

47. Bigourdan, "Les premières réunions savantes de Paris au XVIIe siècle—L'Académie de Montmor," 160, 217.

48. Sturdy, *Science and Social Status*, 26–38. Also see Louis XVI, *Mémoires pour l'instruction du Dauphin*.

49. The monarchy thus went to various lengths to check free thought and religious heterodoxy. See Keohane, *Philosophy and the State*, 262–63. On the perceived threat of libertinage, see Pintard, *Le Libertinage Erudit*, and J. H. Spink, *French Free-Thought from Gassendi to Voltaire*.

50. See also Peter Burke, *The Fabrication of Louis XIV*.

51. Sorbière was not alone in his admiration for Hobbes. On April 25, 1646, Mersenne suggested in a letter that Sorbière examine Hobbes's *De Cive*. Gassendi also noted that *De Cive* was "truly uncommon, and worthy of being handled by all who are sensible of higher things." The letters are printed, in Latin, in the preface to Thomas Hobbes, *De Cive* (Amsterdam, 1647); English translation in the critical edition of *De Cive* by Howard Warrender (1983).

52. *Elémens philosophique du citoyen, où les fondemens de la société civile sont découverts par Th. Hobbes et traduits en français par un de ses amis* (Amsterdam: Blaeu, 1649); *Le corps politique, ou les élemens de la loi morale et civile, avec des réflexions sur la loi de nature, sur les serments, sur les pactes et les diverses sortes de gouvernemens, leurs changemens et leurs relations, par Th. Hobbes* (Leiden: J. & D. Elzevier, 1653).

53. "Discours Scepthique à Philotime, pour montrer que Paris et les Français ne sont pas tout à fait exempts de barbarie, in Lettres et Discours," *Lettres et Discours*, 573–806; "Discours Sceptique en faveur des bêtes et du gouvernement despotique," in *Mémoires de l'abbé de Marolles*, 3 vols. (Amsterdam, 1755), 2: 383–93. Cited and discussed in André Morize, "Samuel Sorbière (1610–70)," *Zeitschrift für französische Sprache und Literatur*, 239–44.

54. Cited in Morize, "Samuel Sorbière," 242. On Sorbière's Hobbesian theory of absolutism, see Keohane, *Philosophy and the State*, 242–44.

55. Sorbière to Hobbes, February 1, 1658, *Lettres et Discours*, letter LXXIX and *Oeuvres de Huygens*, n. 455a, 4: 513–16.

56. Sorbière, *Letters et Discours*, 193–202.

57. See Francois de Graverol, "Mémoires pour la Vie de Messieurs Samuël Sorbiere, et Jean-Baptiste Cotelier . . ." in *Sorberiana, ou les pensées critiques de M. de Sorbiere, recueilliés par M. Graverol, Avocat de la Ville de Nismes. Avec des Memoires pour la Vie de Mess. De Sorbiere et Cortelier*, 2nd edition (Paris: Chez Florentin et Pierre Delaulne, 1695).

58. Vol. One, *Lettres et Discours de M. de Sorbière sur diverses matières curieuses*, appeared early in 1660, and the second volume, *Relations, Lettres, et Discours sur diverses matières curieuses* (Paris, 1660) on October 20.

59. *Lettres et Discours*, dedication.

60. Sorbière to Mazarin, May 8, 1659, *Lettres et Discours*, letter LXXXXXI.

61. In fact, Sorbière and Montmor visited Huygens on April 5 to discuss the reorganization of the Montmor Academy. The project is appended in Huygens's *Oevures Complètes* to a letter written the next day to his brother describing Sorbière's visit. See Huygens to Lodewijk, April 6, 1663, *Oeuvres Complètes*, n. 1104, 4: 323–24, and n. 1105, 4: 325–29.

62. See Sturdy, *Science and Social Status*, 69–71, and Brown, *Scientific Organizations*, 124.

63. Bigourdan. "Les premières réunions savantes de Paris au XVIIe siècle—L'Académie de Montmor," 218.

64. *Sorberiana, sive excerpta ex ore Samuelis Sorbière. Prodeunt ex musaeo Francisci Graverol, . . . Accedunt ejusdem tum epistola de vita et scriptis Samuelis Sorbière et Joan.-Bapt. Cotelier, tum epulae ferales . . . Editio auctior et emendatior* (Tolosae: G.-L. Colomyez, 1694). I cite from the second edition above.

65. See Geoffrey Sutton, *Science for a Polite Society: Gender, Culture, and the Demonstration of the Enlightenment*.

66. *Sorberiana*, 163–64.

67. Sorbière was granted membership in the Society on June 22. Brown, *Scientific Organizations*, 130.

68. Sorbière, *Relation d'un voyage en angleterre, où sont touchées plusieurs choses, qui regardent l'estat des Sciences, et de la Religion, et autres matieres curieuses*, 69, 72.

69. Sorbière, *Relation d'un voyage en angleterre*, 89–90.

70. On the immediate politics behind Colbert's decision to found the academy, see Sturdy, *Science and Social Status*, 63–79; Hahn, *The Anatomy of a Scientific Institution*, and A. Stroup, *A Company of Scientists: Botany, Patronage and Community at the Seventeenth-Century Royal Academy of Sciences*.

71. Huygens to Moray, June 12, 1664, *Oeuvres de Huygens*, n. 1234, 4: 69–70: "L'Académie chez Monsieur de Montmor a pris fin pour jamais, mail il semble que du debris de celle cy il en pourroit ranaistre quelque autre, car j'ay laissè quelques une de ces Messieurs avec de tres bonnes intentions."

72. Sorbière's only remarks about the Chancellor (i.e., Clarendon) appear on page 97 of the *Relation*.

73. Vincent Guilloton, "Autour de la Relation de Voyage de Samuel Sorbière en Angleterre, 1663–1664," 131–32.

Chapter Two

The *Esprits Supérieurs*: Bernard de Fontenelle's Academic Eulogies

Anyone who has read the works of the eminent writer Bernard de Fontenelle (1657–1757) might suppose he rarely felt such dark passions as fear, self-loathing, and despair. Witty and sardonic, his writing dripping with bemused irony, Fontenelle seems too optimistic to have given in to either fear or loathing. Upon his deathbed, his last words, undoubtedly carefully considered in advance, were said to have been, "I feel a certain difficulty of being." As an octogenarian, he is said to have once remarked upon a seeing a beautiful woman stroll by, "Oh, to be seventy again!" Nevertheless, it seems to have been a decisive moment of personal despair that brought Fontenelle to initiate an influential adaptation of both Stoic and Aristotelian ethical thought. Fontenelle's new ethical argument came to fruition when, as permanent secretary of the Royal Academy of Sciences, he articulated a moral philosophy that positioned the physical sciences as the source of virtue, and their practitioners as a new moral elite. Taking up Samuel Sorbière's Neostoic idea that the sciences and savants had a central role to play in reform, and transforming it into something approaching the sublime—a philosophical ethic—Fontenelle invoked savants as the guarantors of social and political tranquility. By lionizing savants so successfully, Fontenelle set the stage for the tragedy of the 1770s in which the philosophical party would seek to join power and reason in perfect unison—and, through hubris, fall.

It was, as I mention above, a moment of personal despair—in particular in his failed early attempt to forge a career as a writer—that convinced Fontenelle to reconsider his moral ideas. But more on that event below. First, let us consider the historical context: Fontenelle lived and wrote in a specific literary and philosophical environment—the world of the classically educated elite, the Republic of Letters, a world of men (and some women) that was traditionally Aristotelian in education but whose Aristotelianism had been sorely challenged by the rise of Neostoic thought in the sixteenth century and by the new sciences of the seventeenth century. An empirical and pragmatic ideal of the physical sciences (promoted by men such as Pierre Gassendi and his disciple, Samuel Sorbière) had emerged in France as part of the intellectual response to the political, social, and religious crises of the sixteenth and seventeenth centuries—in fact, this new ideal was part of the intellectual crisis of the era. In France, this reaction, which historians have traced especially to the development of religious skepticism during the Reformation and Neostoicism during the religious wars, influenced those who either were or supported the goals of the *politiques* (see chapter 1). In turning the quest for knowledge toward utility and probability and away from what they saw as the vain and dogmatic search for certainty, defenders of the new sciences such as Sorbière argued not only that they had an epistemologically valid method but also that this method supported their own social and political desires for tranquility and order under an enlightened central authority. A state-sponsored academy of the sciences, they thought, should help legitimize the government's authority, improving its moral and political credit by associating it with practical scientific innovation and with the expertise of savants.

The proponents of what was, in effect, enlightened despotism did not, however, escape the ultimately pessimistic and elitist view of human nature that characterized the Neostoics and stamped their own moral-political consciousness. The hope that the sciences could help impose order and stability on a chaotic world did not translate into a belief in social revolution. Savants such as Sorbière and other *libertins érudits* believed an egalitarian social model was possible only among *esprits supérieurs* or *esprits forts* such as themselves—that is, among a natural, intellectual elite who could live by laws and regulations they themselves had made. Ideas about the emancipating power of scientific knowledge would eventually emerge among the scientific community in the late eighteenth century (see chapter 5), but from the mid-seventeenth to the mid-eighteenth centuries, intellectual elitism dominated. It was never totally abandoned.

We should be careful, however, not to draw too many connections between the supposedly historically stable honor-ethic of the traditional nobility and the emerging elitist ethic of the scientific community without first coming to grips with the nuances of post-Frondes intellectual elitism in France.[1]

All elitist discourses may have similarities as they emerge from the same general environment and attempt to justify and naturalize claims of privilege and superiority over the masses. The moral idea of the sciences that evolved in and around the Royal Academy of Sciences, however, should be seen as a phenomenon having more to do with Neostoicism and its offshoot, *libertin-*

age érudit, than with the honor-ethic of an increasingly beleaguered warrior and court nobility. By the late seventeenth and eighteenth centuries, the whole notion of nobility in France was in flux, and the old ideas (or parodies of them) served as points against which to define and articulate new ideas.[2]

A coherent moral philosophy of the sciences that fashioned the savant as a moral and political elite began to emerge around the Royal Academy of Sciences at the turn of the eighteenth century. Fontenelle best and most clearly articulated the idea, presenting an early image of what would later become the ideal of the dispassionate, objective, modern scientist and thinker. To understand Fontenelle's moral philosophy in its historical context, however, we should understand its connection to absolute monarchy, Neostoicism, and a critique of noble autonomy and privilege. Positioning the savant as a moral, patriotic, and even meritorious elite, as Fontenelle did, constituted an attack on the perceived values of the traditional nobility and challenged their place in modern society.[3]

During the second half of his life, Fontenelle was at the center of the French scientific community, serving as *secrétaire perpétuel* of the Royal Academy of Sciences from its reinvigoration in 1699 up to his retirement in 1740. During those forty years, Fontenelle was the public voice of the Academy and, arguably, of the sciences in general in France. The moral philosophy of the science that he articulated had radical implications, and he placed such a stamp on the Academy that no succeeding secretary could escape the influence of his ideas or the power of his rhetoric. His impact on the popularity of the sciences and on Enlightenment thought is almost inestimable, even if it has only rarely been acknowledged.[4] The new confidence he helped instill in the scientific community, and his popularization of a new *goût des sciences* that invaded the salons, courts, and literature of Europe, laid the groundwork for a more direct critique the Old Regime social and political economy in the later eighteenth century.

Fontenelle popularized his moral views on savants and the sciences most clearly (although not solely) through a new academy tradition he initiated, the academic eulogy or *éloge*. Although the eulogy was a literary genre adapted from classical antiquity and traditionally applied to dead kings, nobles, and other heroes, we should not conflate Fontenelle's eulogies with those traditionally written for nobles and statesmen. Such a conflation of the "heroes" of the Republic of Letters with noble heroes might be appropriate in earlier panegyrics such as Gassendi's *Vie de Peiresc* (1640): Peiresc was of the old nobility and had distinguished himself as a great lover of learning. Also, the tradition of savants praising their *maecenas* or patrons dates at least to the time of the Emperor Augustus—although Gassendi's *Vie de Peiresc* differed significantly from the classical tradition, since Peiresc was not a patron in the classical sense.[5]

With Fontenelle, the eulogy takes on a different character altogether: savants are praised for the social utility of their researches and for the moral character the sciences engendered in their superior minds. Fontenelle positions these *esprits supérieurs* as the natural elite of French society in contradistinction to the nobility (as traditionally understood). There is no attempt merely to assimilate an intellectual ethos to a presumably more socially acceptable aristocratic one. Even a brief glance at Fontenelle's eulogies reveals an intimate link between natural philosophy, or the natural sciences, and virtue. A more careful examination of his eulogies shows that the link Fontenelle makes between philosophy and virtue was logically commensurate with genuinely held notions of human nature, happiness, reason, and the good. Furthermore, Fontenelle defines these fundamental human notions in the Neostoic tradition. In fact, the relationship in the eulogies between the intellectual elitism of Neostoic thought and the virtues proper to a savant is highly intellectualized. Although they were certainly part of an effort to fashion a favorable public opinion of the academy and its work, the eulogies cannot be seen as an attempt to cultivate nobility in savants by promoting aristocratic modes of decorum and interests in status and reputation over and above intellectual interests.[6]

To understand Fontenelle's *éloges* as an institution in the Academy of Sciences (and in the Republic of Letters in general), we should examine the specifics of Fontenelle's moral philosophy more carefully. On the one hand, the political situation in France and in the Academy of Sciences at the end of the seventeenth century played a critical role in the decision to appoint Fontenelle as secretary. Historians have studied this institutional and political context in detail, so I will only survey their findings briefly here. On the other hand, historians have often overlooked the intellectual context, or rather Fontenelle's own response to the dominant intellectual themes that formed the undercurrent of the Academy and the natural sciences in France: Neostoicism, skepticism, Aristotelianism, and absolute monarchy as a political theory. Thus, I will focus this chapter specifically on a close reading of Fontenelle's moral philosophy in order to explain his articulation of the moral idea of the sciences.

Although not an entirely unchallenged assumption, historians traditionally speak of the decline of the Académie royale des sciences after 1683 and its renewal or renovation in 1699.[7] This story can ultimately be traced to Fontenelle himself, who wrote of the Academy's malaise in his eulogy for the academician Guillaume Hombert, and who wrote the first history of its *renouvellement*.[8] The standard explanation of the Academy's decline, although not one Fontenelle voiced, revolves around the death of its founder, the visionary Minister Jean-Baptiste Colbert, and his replacement as *surintendant des bâtiments*, with jurisdiction over the Academy, by the marquis de

Louvois.[9] Louvois is typically portrayed as narrow-minded, heavy-handed, and ignorant of scientific research. He suffocated the Academy by withdrawing funds, he meddled in its interior functions with poor results, and he undermined the morale of the savants.

This story is not indisputable. The Academy's activities in the 1670s through the 1690s established it as the crown's consultant on technical and military matters and as the arbiter of technological and scientific discovery in France, with the epithet "Approved by the Academy" (*approuvé par l'Académie*) emerging as an inventor's key to success.[10] There is evidence, however, that all was not well in the early decades of the Academy. If it did not decline, it did not grow, either. In fact, it did not seem to do much at all. Morale suffered, pensions fell, expenditure was reined in, and important founding members were lost.[11] Louvois should not take all the blame for this stagnation in the academy; Louis XIV must take on his share of the blame as well.

The monarchy's failure to live up to the ideals Sorbière, Huygens, Perrault, and others in the scientific community had invested in it was an underlying reason for the problems with the Academy. Sorbière had called for the creation of the Academy in the hope that an alliance of the sciences with an enlightened central authority could finally put a stop to unnecessary war and religious conflict and help bring prosperity and stability to France. Louis XIV's abandonment of Richelieu and Mazarin's foreign policies in the 1670s through the 1690s, however, in order to engage in unprovoked and imperialistic wars of aggression, was a blow to such hopes. So, too, were the king's ill-considered renewal of Protestant persecution and revocation of the Edict of Nantes in 1685. Louis XIV's policies placed serious economic burdens on the kingdom, alienated thousands of loyal subjects, and united much of Western Europe against France. In a suffering economy, money that would have gone to the Academy went to the war effort—and it should be noted that Louvois hardly opposed the unnecessary war and was delighted by the vicious persecution of Protestants that ultimately forced many of the most industrious and prosperous to emigrate. Furthermore, Louvois seems to have done little to keep some of the most accomplished (yet Protestant) foreign academicians from leaving France in disillusionment: Ole Christensen Røemer, Gottfried Wilhelm Leibniz, Nicolaas Hartsoeker, and even Huygens all went home and left a demoralized Academy without some of its most accomplished savants. Other committed founding members were gone as well: Perrault died in 1688, and Carcavi in 1684; Auzout left France for Rome in 1685 and never returned.[12] Louis XIV's policies sapped the sciences of funds, vision, and genius and shattered hopes for stability and peace.

The result of the Academy's stagnation was predictable: it accomplished little. During the eight years of Louvois's control, the academy vegetated. Historians have shown that the Academy actually accomplished little in the applied sciences even under Colbert's administration, but it was not idle in the theoretical and mathematical sciences, being endowed with some first-rate geniuses in these fields. Technological and scientific innovation in the eras of both Colbert and Louvois tended to come from outside. The Academy's role was more often to approve and judge the work of others—or at least assert its right to do so—than to innovate on its own. In addition to funding limitations, the Academy was too small to dominate scientific innovation: until 1691 there were only between fifteen and twenty members, and as we have seen, some of the best were gone by the 1680s.[13]

After the death of Louvois in 1691, Louis XIV transferred control of the royal academies to the Minister of State and the Royal Household, Louis Phélypeaux, Comte de Pontchartrain. More committed to the moral idea of the sciences and to their progress, Pontchartrain managed to keep the Academy funded through some of the darkest years of Louis XIV's reign. He even raised the average annual budget by nearly ten thousand lives, although it was still far lower than during Colbert's administration.[14] Major funding was simply not available. Historians have universally regarded Pontchartrain's administration of the Academy as significant, however, for the structural changes it initiated. Under Pontchartrain, the Academy once again found a leader of vision: Pontchartrain's brilliant nephew, Jean-Paul Bignon. Pontchartrain appointed Bignon to the Academy in 1691 to be his official representative, and Bignon was soon elected president. It was an auspicious choice. Bignon immediately began to review the Academy's function and structure. The ultimate result was the new *Réglement* of 1699 that placed the Academy officially under the protection of the king, clarified the duties of members, and did away with the old, Baconian ideal that the academicians should publish their works anonymously—a practice that stifled innovation by denying credit to individual savants for their work. The new regulations also increased the number of academicians. Between 1691 and 1699, Bignon had twenty new academicians elected, including such famous names as Guillaume Hombert, Joseph Tournefort, and Jacques Cassini.[15] Overall, membership was increased to fifty academicians, plus twenty adjunct positions.[16] In 1697 Bignon also brought his friend, Fontenelle, into the Academy as the new *sécretaire pérpetuel.*

Fontenelle, as we shall see, had already made a name for himself as an interpreter of the sciences for the reading public, and his belief that philosophy and the sciences had a public role of both pragmatic and moral proportions had apparently convinced Bignon. In his effort to revive the Academy and improve its image—and therefore enhance its prestige and funding—Bignon relied heavily on Fontenelle. Accordingly, Bignon opened to the public the annual Academy assemblies, in which Fontenelle began to read his eulogies of deceased academicians after the savants had read their scientific papers. Bignon also put Fontenelle in charge of the new, annually published

Histoires et mémoires de l'Académie royale des sciences, which was soon to become the most important scientific publication in France. When Pontchartrain appointed Bignon as "director of the book trade" in 1699, he used his position to encourage sound scholarship—that is, practical work that did not undermine the authority of the monarchy—and he brought along Fontenelle as one of his censors.[17] In 1700 Bignon also took over the financially ailing scholarly journal, the *Journal des sçavans*, and turned it into an official outlet for scientific publications and reviews. Six special editors were paid to analyze all submissions. Fontenelle, who had some skill in the field, took charge of those on mathematics.[18]

This, then, was the political and institutional context in which Fontenelle began to outline a moral philosophy of the sciences in the *éloges*. Of course, these institutional imperatives did not determine what he wrote. Although Fontenelle carried out the duties Bignon assigned to him, one should not assume that he was merely Bignon's mouthpiece, or that his goal was limited to improving the image of the Academy. Fontenelle was not a propagandist. The eulogies expressed a coherent moral philosophy of the sciences that can best be understood if we consider the depth and seriousness of Fontenelle's thinking as he responded to the intellectual currents and crises of his time.

Although it is possible to see the institution of the academic eulogy within a larger context of French eulogy writing that was concerned with moral instruction—providing stimulating examples of *politesse*, for example—the academic eulogies are best understood as part of the neoclassical tradition (of which Neostoicism was a central part) that so dominated educated culture in France. By the time of Fontenelle, neoclassicism in general was a dominant fashion in art, architecture, drama, and literature. Although fashionable ever since the Italian Renaissance had begun to influence French culture in the fifteenth century, the promotion of neoclassical modes became a veritable state program during the reign of Louis XIV. By the late seventeenth century, the monarchy had mobilized the arts as never before for the display of the *gloire* of the Sun King. The monarchy actively promoted classical models as expressive of the order and harmony it wanted to symbolize. Many savants, who, as we have seen, emerged from a Neostoic intellectual tradition that longed for order and harmony above all else, fully agreed with the neoclassical ideals and supported the idea of an absolute monarchy in which sovereignty was undivided. It is therefore no accident that the classical elegiac tradition found its way into the Academy of Sciences.[19] The *éloges* in general, and Fontenelle's in particular, were a part of the neoclassical enterprise.

The institution of the *éloges* in the Academy, which began with Fontenelle and evolved in the hands of his three successors as *secrétaire perpétuel* (d'Ortous de Mairan, Grandjean de Fouchy, and the marquis de Condorcet), emerged from the panegyrical rhetoric of classical antiquity. Orators of the ancient Greco-Roman world engaged in panegyric to represent the virtuous behavior of dead heroes so the living had models for emulation. The panegyric was, thus, a rhetorical and oratorical practice with mimetic and ethical, but not primarily historical, intent. Even though eulogies related the lives of heroes, they were not biographical in the critical sense, as we might understand it today. Furthermore, the eulogist was limited to a stock of *topoi*, or a class of topical arguments that could illustrate the ethical purpose of his rhetoric. For instance, he might use a type of mock humility in the formal exordium of the eulogy in order to convince his audience of the particular worthiness of his subject.[20] In other words, panegyric turned away or "troped" from "realistic" discourse (in the sense of following formal rules of logic) and found refuge in certain tropes, or moods, or figures of speech.[21] The purpose of the panegyric was to be useful to the present, not to be a true representation of the past; its auditors understood and accepted this fact while still enjoying the moral truth contained in the rhetoric.

Like many of his contemporary men of letters, Fontenelle would have learned the classical tradition of panegyric and its accepted *topoi* as part of his study of rhetoric at his Jesuit *collège*.[22] As a result, Fontenelle's notions of how and what virtues to praise was derived directly from the classical rhetoricians, such as Plutarch, Cicero, Quintillian, and the rest. As much as Fontenelle was steeped in the classical tradition, however, he was acutely aware of the differences between his time and antiquity. In accordance with what he thought was appropriate for his own day, he had no qualms about diverging from the classical panegyrical tradition and striking a greater balance between a biographical and ethical purpose for the *éloge*. His eulogies often went into great detail on the life and works of each savant. Nevertheless, he always kept sight of the ultimate goal: to idealize the virtue of the savants he eulogized.

Scholars examining the *éloges* have linked Fontenelle's expressed moral philosophy to those virtues appropriate to the classical panegyrical tradition, especially those expressed in Plutarch's popular *Parallel Lives*. In other words, the Roman Stoic virtues that Plutarch praised in the heroes of antiquity are the same ones that Fontenelle praised in the "Plutarchian heroes" of the seventeenth- and eighteenth-century scientific community. Just like Plutarch's Coriolanus, Marcus Cato, and Sertorius, Fontenelle's savants embodied Stoic fortitude, and like Pericles, Solon, and Lycurgus, they possessed an overwhelming sense of duty. Perhaps most importantly, however, they expressed "temperance and equanimity" and "courage and resolution"; they were endowed with the pastoral, Arcadian virtues of "simplicity, humility, honesty, want of ambition, poverty, austerity, and frugality"—all, of course, virtues of particular value to Roman Stoicism.[23]

The ancient heroes may have exhibited these virtues, too, but it is because they are Stoic virtues that Fontenelle consciously incorporated them into his image of the functioning modern savant. Fontenelle's adaptation of the Stoic virtues must be seen as something more than mere worship of classical modes or the colonization of his mind by a neoclassical discourse. Fontenelle's *éloges* celebrate an Arcadian goodness, associated with his own mythic notions of the shepherd's life, a life uncorrupted by civilization and commerce with humanity.[24] In Fontenelle's eulogies, the sciences enabled an *esprit supérieur* to achieve the Stoic state of *apatheia* that freed him from immoderate passion and desire, and thus engendered virtues such as simplicity, humility, honesty, austerity, frugality, and want of ambition. Fontenelle characterized this condition in his own language as *tranquillité*.

For instance, the affable and modest academician Vincenzo Viviani "had an innocence and simplicity of manners that one finds ordinarily in those who have less commerce with men than with books."[25] Guillaume de Amontons had a "simplicity, a frankness, and a candor that a minimum of commerce with men can conserve."[26] Natural philosophy somehow engendered an Arcadian freedom from the corruption of modern life. In so doing, it enabled men to overcome the violent passions that governed the lives of most and achieve a true tranquility of soul. Indeed, the manners of Pierre-Sylvain de Regis were "such as the study of philosophy can form, when it does not find too much resistance from the nature of the man."[27] As for Ernfroi Walter de Tschirnhaus, "The true philosophy had penetrated into his heart, and had established there that delicious tranquility which is the greatest but the least sought of all goods."[28] Likewise, Guillaume Homberg had achieved "that tranquility of soul" so that he was "free of the tumult of the passions." In fact, "A sound and peaceful philosophy had disposed him to accept without trouble the different events of his life, and it rendered him incapable of those agitations to which we are so subjected."[29]

Even those with whom Fontenelle did not agree were worthy of tranquility if they devoted their lives to natural philosophy. Fontenelle at the time championed Cartesian physics; yet Isaac Newton still possessed, according to Fontenelle, the virtues of modesty and humility "augmented by the wise simplicity of his life."[30] The tranquil geometer Pierre de Varignon had a character "as simple as his superiority of mind was able to ask." Fontenelle was very clear that it was natural philosophy itself that induced receptive men to the good: "I have already given this same praise to so many persons in this academy that one might believe this merit pertains more to our sciences than to our savants."[31] Those who practiced natural philosophy were by that very act elect; of the elect, those who attained a life of perfect contemplative virtue achieved a philosophical sainthood. Regarding Jean-Baptiste du Hamel, his predecessor as *secrétaire perpétuel*, Fontenelle found himself (rhetorically) at a loss for words. He could not express the virtues of the man "because this would be the panegyric of a saint." Nevertheless, he quickly recovered his composure and remarked, "One saw easily that his humility was not mere acting, but a sentiment founded on science itself."[32]

Fontenelle expressed what he idealized as the innate virtues of savants with a literary grace hardly revealed in the brief excerpts I have translated here. What is more, a unified vision of the intellectual virtues runs throughout his eulogies. The virtues Fontenelle praised may have been linked to the Plutarchian ideals necessitated by the *topoi* of the classical panegyrical tradition, but they were not fully determined by them. Fontenelle's notion of the intellectual virtues, as articulated in the eulogies, are not merely bound up in Plutarchian tropes, and the pastoral aspect of the virtues requires more attention. We should be careful not to overcontextualize Fontenelle's ideas, thus freezing them within their own history and preventing them from saying anything new. For Fontenelle, the savant exists in the realm of truth, having overcome, not denied, the woes of earthly existence by perfecting his own human nature. There may be a personal psychology at work here, but it is one with a clear understanding of the Platonic-Aristotelian philosophical tradition as well as with the hugely popular Stoicism as interpreted in sixteenth- and seventeenth-century France.[33]

Fontenelle was not a systematic philosopher. He wrote no unified philosophy of ethics, or epistemology, or metaphysics. Fontenelle was a *littérateur*, and what he wrote was mostly literature. This does not mean, of course, that he did not have a philosophy, or that he was incapable of engaging in the philosophical issues of his day. It was precisely his ability to discuss philosophy in a clear, engaging, literary style that earned him the patronage of Jean-Paul Bignon, the admiration of the academicians, and eventually the position of *secrétaire perpétuel* in the Académie royale des sciences. Who better to write the history of the Academy than one of the foremost literary defenders of the sciences of his day?

Fontenelle developed his notions of human nature at a fairly young age, beginning in his mid-twenties, and they remained surprisingly constant throughout his life. In short, his philosophy of human nature is pessimistic, but it is a mitigated pessimism. A historian once cleverly named this philosophy *naturalistic fatalism*, but such philosophic jargon should not suggest that Fontenelle came to his conclusions in pure, contemplative isolation from life.[34] In works such as the *Nouveaux dialogues des morts*, "Discours sur la patience," and "Du Bonheur," we can see him wrestling with and working through personal despair: Why am I a failure? Why am I unhappy? What is the meaning of my life? His philosophy is embedded in a seventeenth-century literate context and is heavily engaged with sixteenth- and seventeenth-century Neostoicism—although no mere copy of it—but in the end he is asking and answering real, personal questions about his life. We may fairly speculate that his philosophy, or rather the disaffected angst only partially submerged beneath it, began to emerge, perhaps first as inchoate feeling, in his formative years, when he was Bernard *le raté*—Bernard the failure.

Fontenelle's father, François, was a lawyer at the Parlement of Rouen, and so it seemed practically preordained that young Bernard would follow him into the vocation. Fontenelle studied law and rhetoric at the Jesuit Collège in Rouen, and by 1674, at seventeen years of age, he was accepted to the bar. He pleaded his first case that year—and lost. Of the possible mitigating circumstances, we have no idea. According to some interpretations, Fontenelle had never wanted to be a lawyer anyway and did not get along with his father, who was reputed to be of little intelligence and "boorish humor."[34] According to other stories, the court was either struck by Fontenelle's eloquence or, on the contrary, his speech faltered, which seems more likely.[35] At any rate, Fontenelle renounced the bar after just this one case. Not surprisingly, he soon found himself out of his father's house and on the road to Paris—apparently to try his luck as a man of letters. He did not have the means to remain in Paris on a permanent basis until the late 1680s, but from 1675 onward he spent as much time in the city as possible.

Fortunately for Fontenelle, his mother's brothers were the famous Corneilles: Pierre the dramatist, and Thomas, editor of the literary journal *Mercure galant*. In Paris, Fontenelle rubbed elbows with the literary elite, the libertines, and the *érudits*, and he was a favorite of the *précieuses*, the hostesses of the salons. More importantly, Fontenelle had the *Mercure galant* as a literary outlet. Normally authors had to pay to have their works published in the *Mercure*; Fontenelle, of course, did not. Throughout the 1670s and 1680s, his poetry and literary pieces appeared regularly in the journal. This is not to say he had achieved success on his own merits. In 1676 he submitted a piece for the poetry prize at the Académie Française—and lost, once again. (Consequently, the Academy became a particular obsession with him: he tried and failed four times to be admitted before finally being accepted in 1691. Of course, by that time his uncle Thomas was the chancellor of the Academy.) Perhaps worse, his first attempt at a tragedy, *l'Aspar* (1680), flopped miserably, and Racine himself heaped criticism upon it. Humiliated and despairing of success, young Bernard burned the manuscript, along with the last of his optimism, and retreated to Rouen.[36]

It was therefore an irony undoubtedly not lost on the intelligent and still young man (he was twenty-six) that he finally achieved real literary recognition for a work written in self-imposed exile back in Rouen, a work that revealed a deep, satirical pessimism about the human condition, the *Nouveaux dialogues des morts* (1683). Modeled on the original second-century *Dialogues of the Dead*, Fontenelle took the role of a new Lucian. Through the mouths of the famous dead, Fontenelle satirized that which (to paraphrase Melville) most maddened and tormented him—all truth with malice in it, all the subtle demonisms of life and thought. When he proclaimed that humans were selfish and stupid and governed by their passions, he was speaking of himself, too. It was his ambitions, his passions, and his stupidity that had led him to abandon the practice of law and set him on a course for seeming failure as a man of letters. Therefore, men were fools if they thought reason could achieve dominance over the passions.[37] Perhaps this was what he had learned from his one case as a lawyer: in the courts one could overcome the passions with reasoned arguments, and yet his own human nature, his fear and lack of self-confidence, had caused him to stammer and falter—and fail. Humans, he saw, are essentially passionate animals, and it was upon this fundamental premise of human nature—the exact opposite of Aristotle's—that Fontenelle built his ethics, using Aristotle's own logic.

In the guise of Anacreon, the Dionysian lyric poet, he therefore mocked Aristotle's dictum "Man is man by reason alone." Rebuffing Aristotle's patronizing banter concerning his interest in wine and song, Anacreon replied in Stoic terms, "It is more difficult to drink and sing as I have sung and I have drunk than to philosophize as you have philosophized. For to sing and drink as I, it would be necessary to have cleared your soul of violent passions, no longer to aspire to that which does not depend on us, to be disposed to take each day as it comes."[38]

Fontenelle saw that it was passion, especially aspiration, that led to unhappiness (or at least his own). If one could but follow the advice of Guillaume du Vair and the Neostoics and maintain apathy toward the different events of life, then one could achieve happiness. But nothing could be more difficult; even the metaphysical philosophizing of Aristotle pales in comparison. This is because man is man by passion, with perhaps a smattering of reason in passion's service. In fact, even as he admired the Stoics' admonition to disavow the desires, Fontenelle was not sure such a disavowal could be possible or for the good: "It is the passions that make and unmake all," he wrote. "If reason dominated the earth, nothing would happen. They say that pilots fear passive seas where they are not able to sail most of all and that they want wind, even at the risk of a tempest. The passions are for men as the winds that are necessary to put everything in motion, even if they often cause storms."[39]

In fact, Fontenelle began to presage Nietzsche in his claim that the apotheosis of reason might kill life: "She [Nature] has put men on the world to live, and to live means not knowing what we do most of the time. When we discover the small importance of what we do and what concerns us, we steal from nature her secret: then we become too wise and no longer want to act, and this nature does not find good."[40] In other words, humans must have the capacity to live irrationally. We must believe that what we do has meaning or that things are as they seem even if our reason tell us otherwise—or else we risk a crisis of being that would freeze us in despair.

By defining man as a creature of passion, and thus rejecting the Cartesian view of man as a thinking thing, as well as the Aristotelian view of man as a rational creature, Fontenelle mitigated his despair and absolved himself of his failures. He was bound to fail because he was bound to desire and to aspire. In his view (admittedly hardly an original one), passion became the motor of history, the cause of human actions, both evil and good: "All great deeds which men were compelled to do were done without reason. The order that nature seeks is always enforced."[41] To deny one's essential nature, then, is to act unnaturally—in Stoic terminology, not to be in accordance with nature, and therefore not to be virtuous. In commonly understood Stoicism, happiness (or at least Stoic indifference, or apathy) results only from being in harmony with nature. Not coincidentally, then, happiness is the second key element underlying Fontenelle's ethics. Like Aristotle, Fontenelle understands happiness as the greatest human good and virtuous or good behavior as necessarily in accordance with happiness.

In redefining human nature, however, Fontenelle had dug himself into a hole. If to act in accordance with nature meant surrendering ourselves to the passions, then we are bound forever to desire things "not up to us"—and thus to be cursed to unhappiness. It is for this reason, I think, that the *Dialogues* never arrive at a truly positive closure. Fontenelle would eventually find his way out of this dilemma by reinterpreting Aristotle's ethics in light of his own view of human nature. This interpretation would, as it had for Aristotle, make philosophy—or, rather, for Fontenelle, natural philosophy—that activity that induces virtue and happiness. But in 1683 Fontenelle had not yet built this intellectual bridge, and his pessimism dominated. What hope is there for man, he despaired, doomed to unhappiness and kept in ignorance by the very nature of his being? In the guise of Montaigne, Fontenelle laughed at Socrates himself for believing men might have ever improved their nature: "Men of all centuries have the same proclivities, over which reason has no power at all. Thus, everywhere there are men, there are follies, and they are the same follies."[42]

Fontenelle was perhaps never so harsh on human nature in the *Dialogues* than when he spoke through the character of Molière. When Paracelsus asked him to what use he had put his study of human folly, Molière quipped, "I assembled in a certain place the greatest number of people that I could, and there I made them see that they were all fools." Amazed, Paracelsus wondered how Molière could possibly persuade them of such a thing, to which Molière replied, "Nothing is easier. One proves their folly to them without employing any great feats of eloquence, or any well-considered reasoning. What they do is so ridiculous that it is necessary only to do as much before them, and you will soon see them burst out laughing."[43] Thus, by their very nature humans are prone to folly and foolish behavior. In all his satire, however, Fontenelle revealed here his great debt to Molière. If anywhere, this was where he explained the point of the *Dialogues*—a point that is in its essence *comédie molièrenne*: "In order to laugh at the world," Fontenelle explained, "it is necessary to be in some fashion outside of it, and comedy draws you from it: it gives you all of it as a spectacle, as if you had no part in it at all."[44] The *Dialogues* are cathartic; that is the source of their charm, and probably why they sold so well. In a mad world, what saner response can there be than comedy? In explaining what humans are and laughing at their idiocy, Fontenelle had perhaps to some degree put himself beyond humanity, and thus he could feel he had escaped his own, rather minor, troubles for a while. Even if the *Dialogues* ultimately offered no answers, the laughter was good medicine.

Being a series of satirical vignettes and not a sustained argument, the *Dialogues* say many things on different subjects. Reading them in a unified manner risks an interpretation that perhaps could be questioned with evidence internal to the text.[45] But the tone of the *Dialogues* seems to me too harsh, at times too shrill, merely to be a disconnected literary effort aping the libertine, skeptical, and Neostoic sensibilities of a Molière, Cyrano de Bergerac, or even Montaigne.[46] Fontenelle himself recognized the irony of the *Dialogues*' success—that is, the literary success of a work in which he explores and rails against the underlying reasons for his own failure as a young man of letters. Possibly with an ironic grin (the one we see in Hyacinthe Rigaud's famous portrait of him?), Fontenelle himself anonymously published a critique of the *Dialogues* the year after they appeared, titled *Jugement de Pluton*, a final joke on humanity as a whole, this time from the lips of the god of the underworld himself.

Although Fontenelle's philosophy of human nature began to form in the cathartic outpouring of 1683, no specific ethic had yet emerged. Nor could it. In the *Dialogues* the moral order of the cosmos is thrown on its head. Good is bad, and bad is good: to be in accordance with nature (that is, to be virtuous) is to be a passionate animal, but that precludes happiness. According to the Aristotelian tradition, happiness is a disposition of the soul in accordance with goodness. Thus, if one is not happy, one is not virtuous, or not disposed toward the good. This confused situation possibly mirrored Fontenelle's fear and self-loathing. It was in his attempt to find happiness that Fontenelle began to rework the Neostoic ethic of intellectual elitism that placed natural philosophy at the center.

After the *Dialogues des morts*, Fontenelle's luck slowly began to change. He learned that, despite his love of poetry and drama, his real literary gift was for satire, comedy, and philosophy. He was at his best when seriously considering natural philosophy in a satirical, comedic tone. Not long after the *Dialogues*, his interests turned toward the new sciences, and he published the work that would bring him lasting fame, the *Entretiens sur la pluralité des mondes* (1686). Certainly influenced by the work of the *libertins érudits* of Paris, and probably by Cyrano de Bergerac's satirical *l'Histoire comique des états et empires de la lune* (1657) and *l'Histoire comique des états et empires du soleil* (1662), Fontenelle earnestly considered the possibility of life on other planets, but always with an eye to comedic commentary on the sciences and philosophy. Cast in the form of a dialogue with a young marquise, Fontenelle remarked wryly in the preface, "Don't you believe that if Wisdom wished to present herself successfully to men, she would do well to take a form such like that of the Marquise? Indeed, if Wisdom could make her conversation equally agreeable, I assure you that all the world would run after her." Despite his newfound literary success, Fontenelle's view of human nature had not changed: humans are governed by their passions, and men would much rather chase after beautiful women than truth itself. (Fontenelle had a reputation as a philanderer, so here again he was laughing at himself.) Natural philosophy, where it exists at all, must therefore be a result of our passions and our ignorance: "All philosophy . . . is based on two things only: curiosity and poor eyesight; if you had better eyesight you could see [the universe] perfectly . . . and if you were less curious you wouldn't care about knowing, which amounts to the same thing."[47] In general the *Conversations on the Plurality of Worlds* revealed the persistence of Fontenelle's notions of human nature as first elaborated in the *Dialogues*.[48]

We have seen that the *Dialogues* end rather sourly with Fontenelle finding catharsis in laughter, but at the cost of having defined happiness (true happiness—pleasure is another matter) out of the nature of man. That is, happiness became a state beyond the grasp of mortals. To be in accordance with nature, as Stoicism requires, means acknowledging that we are passionate creatures, but by that acknowledgment we realize we cannot, as the Stoics suggest, "despise what is not up to us"—in fact, we cannot help but desire it. Thus, to be as we are means being cursed to unhappiness. This Fontenelle found insupportable. An initially unpublished essay titled "Du Bonheur" can be read as Fontenelle's attempt to climb out of the philosophical hole he had dug for himself. How could he retain his understanding of human nature—a nature he saw confirmed everywhere he looked, especially when he looked inward—and yet still salvage some hope of happiness in life?

"Du Bonheur" unfortunately sends conflicting messages. Fontenelle does not seem to be sure at times what he wants to do with it (whether it should be a theoretical analysis or a practical guide) or how he wants to define happiness (as a state of being or as a way of experiencing pleasure). Perhaps this is why he did not publish the essay immediately, even though it was written fairly early, most likely sometime in the late 1680s or around 1690 to 1691.[49] The work is interesting, however, because it paradoxically marked a further break from, yet at the same time was a confirmation of, Neostoic principles. Concerning Neostoicism, Fontenelle repeated what he had argued in more depth in the 1687 essay "Discours sur la patience"—namely, that he rejected what he saw as the Neostoic and Cartesian belief in a duality of mind and body. Humans are passionate animals, and one cannot imagine that their reason or their minds are separate from their bodies. Fontenelle, therefore, also rejected the Neostoic notion that humans could disavow desire or ignore the evils that plagued them by claiming these evils had no power to affect their true being (their minds): "When a Stoic, weighed down by the sadness of a violent malady, cries out: *I will not avow however that you are an evil*; the effort that he would make to disavow it, is this not actually the strongest and sincerest avowal that can be made?"[50] Yet Fontenelle continued to believe, as the Neostoics did, that humans could, to a certain degree, control their judgments of the world, and this enabled them to establish some control over their passions.

"Du Bonheur" finally defined happiness as a state of being, "a situation such that one desires it to last without changing."[51] It is not the same thing as pleasure, which is merely an agreeable sensation that quickly passes. Nor is it simply caused by pleasures, since some people may experience pleasures and never find happiness, while others live plain lives and yet still enjoy happiness. True happiness is a state of psychological *immobility*, or (to use the Neostoic term Justus Lipsius made famous) *constancy*. The only concern of he or she who achieves this state is to conserve it.

The Stoics, explained Fontenelle, say that one can achieve this state by eliminating desires, by not wanting things to be other than they are. This is folly, Fontenelle argued, for it requires us to maintain "the ridiculous and useless vanity that we are invulnerable."[52] That is, it stipulates that we are essentially thinking things that only happen to have bodies under our complete control. If happiness exists at all, and it is not the result of an elimination of or a total abandonment to the passions, then it necessarily must find itself in a moderation of the passions. Here Fontenelle began to approach the elitist (even aristocratic in the Aristotelian sense) notions we find in his academic eulogies that link him philosophically to Pierre Charron and the *libertins érudits*: most people are of such a character and disposition that they will never be happy, but there are a few "calm and moderate" people "who accept more willingly agreeable ideas and impressions: these people can work usefully to make themselves happy."[53] They are ultimately the good, virtuous, and philosophic people, and thus, by extension, an intellectual elite—the *esprits supérieurs*.

In order to moderate the passions and approach happiness, one must, according to Fontenelle, first cleanse the soul "and chase away all evil fantasies."[54] That is, through some sort of philosophical contemplation it is possible to distinguish between what we imagine to be misfortune or what we presume deserves our desires and passions (which, however, are not real at all) and what truly deserves our attention. People often imagine things to be worse than they are, or they fail to recognize the good they have, always desiring what are falsely perceived as greater and better goods: "they do not deign to stop and taste those [goods] that they have; often they abandon them in order to run after those they do not have."[55] In other words, all that we could desire to make us happy, the "small goods," are cheap and readily available, but many, opting for immoderation, allow desire for perceived greater goods to govern them, and thus they find only despair. The important assumption here is that the object of all desire is the good, and it is the proper perception of what is truly good that brings happiness, not the fleeting pleasures of the false goods.

At this point Fontenelle critiqued what he saw as the decadence of the courtly life, dominated by the false values of *esprits mediocres*, and he denied it was a true good, a proper object of desire. If we examine the perceived greater goods, he argued, the objects of our violent desires, we often find they melt away under the illumination of reason: "Why is this status that I pursue so necessary to me? It is because I must be elevated above other men. But why is this necessary? It is to receive respect and homage from others. But what do I get from this respect and homage? Men will flatter me very noticeably. But how will they flatter me, when I owe the flattery only to my status, and not to myself?" Fontenelle concludes, "Often our idea of happiness is much too complex. How many things, for example, would be necessary in order to make a courtier happy? The credit of the ministers, the favor of the King, considerable establishments for him and his children, luck at games, faithful mistresses who flatter his vanity; finally everything which could represent to him an imagination unrestrained and insatiable?"[56]

It is a matter of calculation, argued Fontenelle: How much do we value those pleasures only gained through immoderate desires, and how do we weigh them against the pain we must endure in order to obtain them, and against their instability in time? How are they worth it when the simple pleasures, "tranquility of life, association, hunting, reading, etc. . . . will always be pleasurable, and cost nothing? Immoderate men might find these simple pleasures insipid . . . but what they call insipid, I call tranquility . . . what idea do they have of the human condition when they complain of being only tranquil? It is the most delightful state that I can imagine."[57]

An ethic is emerging here based on a particular conception of human nature and humanity's proper orientation toward the good. It is elitist in the Neostoic sense, but it is highly critical of traditional notions of happiness and success in Old Regime elite culture. The good is not the achievement of *being through recognition* in court society—according to whatever criteria that society establishes as necessary for recognition, be it *politesse* and *honnêteté* or something else—but it is individualistic: it is whatever allows one to be well and whole within one's self. The object of our desires and passions should be that which brings us the inner tranquility of being sovereign individuals. Fontenelle advocated a retreat into the self, as had Lipsiüs and Montaigne, for example, and he maintained the Neostoic ideal of "moderation" as a supreme virtue.[58] The lessons of the religious and civil wars were not forgotten.

Fontenelle was well aware that he articulated an ethical position in "Du Bonheur"—the notion of happiness has rested on an ethical foundation at least since Aristotle. As the essay drew to a close, Fontenelle wrote, "The greatest secret to happiness is to be well with one's self. Naturally all the troublesome accidents that come from outside throw us back into ourselves. It is good to have an agreeable retreat, but the self is only able to be so if it is prepared by the hand of virtue."[59] Virtue in this ethic, however, was naturally founded on notions of a sovereign self. We must not be ashamed of the inner self, argued Fontenelle, when we retreat there for reflection and contemplation—sociability was not the issue. Virtue is conducive to neither riches nor social elevation, Fontenelle wrote, but brings "an infallible compensation: interior satisfaction." One therefore "finds in one's own reason and integrity the greatest depths of happiness for which others hope in the caprice of chance."[60]

It is becoming clear that Fontenelle's attempt to achieve happiness within a particular notion of human nature involved the articulation of an individualistic ethic that owed much to, but was no mere copy of, the Neostoic tradition. Fontenelle's ethic differed from the Neostoicism of Lipsius or du Vair in its insistence on the primacy of the passions, but it adapted many of the Neostoic virtues. And what are these virtues? They are moderation, simplicity, humility, and a focus on the sovereign self and its own tranquility—in other words, exactly those virtues Fontenelle praises savants for exhibiting in his *éloges*.

At least by the time he became secretary of the Académie royale des sciences, Fontenelle had decided that it was natural philosophy itself that induces these virtues in *esprits supérieurs*, leading through the sciences to a proper recognition of the good. Fontenelle's notion of the good consisted in a modern, Neostoic, and individualistic orientation inward, but the role he gave natural philosophy drew on ancient traditions perhaps inaugurated by Plato and articulated most clearly by Aristotle. In fact, "Du Bonheur" followed many of the arguments of the *Nichomachean Ethics* with striking similarity. The significant difference was in the definition of human nature.

A brief digression on Aristotle's *Nichomachean Ethics* reveals the parallels and divergences of Aristotle's and Fontenelle's moral views, and it demonstrates why the sciences play a central role in inducing virtue for Fontenelle. In the *Nichomachean Ethics* Aristotle defined the *telos* of human life as happiness itself: "What is the highest of all practical goods? Well, so far as the name goes there is pretty general agreement. 'It is happiness'" (I.iv). Like

Fontenelle, Aristotle argued that happiness is not the same thing as pleasure. It is "an activity [or disposition] of the soul in accordance with virtue," by which he meant the "best and most perfect kind" of virtue. Note that *virtue* is a translation of the Greek *aretē* (excellence), which denotes the abstract noun for *agathos*, or "good."[61] So happiness consists of acting or being in accordance with excellence or goodness. Both Aristotle and Fontenelle rejected the Platonic notion that there is a universal Good. Aristotle argued merely that there is a "good for man" (meaning all men) that is the ultimate object of human life. Fontenelle rejected the whole notion of universals and maintained a much more pragmatic view (see below). Aristotle linked this human good to his notion of human nature: man is a rational animal (but remember, he meant something different from a Cartesian "thinking thing"). Therefore, to be in accordance with virtue or goodness is to be perfectly rational. Fontenelle, of course, rejected this notion of human nature, but his logic was the same as Aristotle's. For Fontenelle, man is a passionate animal, and so to be in accordance with virtue, or goodness, is to be perfectly passionate—that is, not to give in to wild, violent, uncontrollable passions but to be properly passionate toward, or desirous of, what is truly good, and therefore conducive to tranquility and happiness. As Aristotle's man is led toward virtue by his study of philosophy, so, too, is Fontenelle's—but it is through natural philosophy.

For Aristotle, the average good man is one who bases actions on reason. Here he meant reason in the practical, calculative sense of choosing well in the everyday activities of life. The man of perfect goodness, and therefore of the highest moral character, however, is more than this. In book VI, the "Intellectual Virtues," Aristotle argued that there are two distinct halves to the rational part of our souls: the "calculative" discussed above, and a higher "scientific" part (VI.i.). The calculative part deliberates on that which is variable, but the scientific part contemplates the invariable—that is, the universal and unchanging truth of the world. Thus the "best and most perfect" kind of virtue, according to which the soul must be disposed to achieve the highest happiness, is scientific contemplation, which for Aristotle is true philosophy (X.vii.). Because philosophy is the apprehension of the "highest things that can be known," its aim is nothing other than contemplation for its own sake. Hence it "entails pleasures that are marvelous in purity and permanence," much like Fontenelle's "simple pleasures" (X.vii.). We must remember here that Aristotle conceived of this higher form of reason not as a procedure for use in seeking knowledge of truth but as the perception of truth itself. You cannot be reasonable in the Aristotelian sense and still be wrong about how things really are. Thus contemplation of truth has no goal other than the pleasure of pure contemplation. The best and most virtuous man is the happiest man because he is perfectly wise. Moreover, he is a "self-sufficient" man because he needs no others to confirm his virtuosity: "the wise man can practice contemplation by himself, and the more wise he is, the more he can do it" (X.vii.).

For Fontenelle, philosophy as a contemplative activity played the same role it did for Aristotle, but in accordance with his diametrically opposed view of human nature and a different notion of knowledge. In an unpublished fragment on human reason, Fontenelle rejected the notion of universals, arguing that they are merely linguistic constructs. Like the *libertins érudits* and most of the scientific community, Fontenelle supported a pragmatic, empirical epistemology: sensible knowledge of the particulars of the world was possible.[62] Thus, for Fontenelle, it was through the contemplative search for the practical and useful truths (and thus the practical and useful goods) of the world that the philosophical mind finds the virtue to direct his natural human desires toward those objects proper for ensuring his happiness. We are always impassioned by desire, but the sciences induce us to desire those simple, permanent goods that bring tranquility. As he wrote in his eulogy of Homberg, "Whosoever has the leisure to think discovers nothing better than to be virtuous."[63]

Ergo, for Fontenelle the most virtuous men in his age, those heroes most worthy of eulogies, were savants and natural philosophers. Far from promoting any sort of mimicry of the grasping court nobility, Fontenelle all but denied this nobility any possibility of discovering what the *esprits supérieurs* of the natural sciences had discovered: true happiness, tranquility, and the good life. Moral superiority rests with the intellectual elite, which, through rational, practical, scientific inquiry, benefits both itself and its society.

In closing this chapter, I want to say something about the pastoral element in Fontenelle's philosophy, a seeming contradiction to a philosophy of progress, science, and modernity. Like many of his contemporaries at the end of the seventeenth and beginning of the eighteenth centuries, Fontenelle mused considerably on a mythic past for examples of moral purity useful in the present. Such musings must be differentiated from the primitivism of those, such as Rousseau, who would later reject the ideal of moral and technological progress imputed to the arts and sciences and who seemed to suggest the desirability, but impossibility, of a return to some past era of genuine virtue. In various writings, Fontenelle displayed his hostility to the "decline and fall" version of history, and he refused to imbue the ancients with any greater degree of virtue than the moderns.[64] Fontenelle's fascination with Arcadian pastoralism was an attempt to locate a past moral example that highlighted his own belief in the classical Stoic, temperate, and orderly virtues.

Unlike the robe and sword nobility that sought, through the arguments of moralists such as Boulainvilliers and Montesquieu, to justify its authority and its vision of moral reform by appealing to a pseudo-historical, feudal, Germanic constitution, Fontenelle appealed to a classical, Arcadian myth.[65] In the Arcadian myth, the shepherd, not the warrior, is the moral example because his life is orderly, simple, and genuine, not heroic, martial, or severe. In Fontenelle's musings on pastoral poetry, for example, we can see him beginning to work out his ethics before they came together more clearly in "Du Bonheur." Already the crucial ideas—tranquility, simplicity, and the rejection of the immoderate courtly life—are there. If read in light of his 1687 essay "Discours sur la nature de l'églogue," one sees that human happiness as defined in "Du Bonheur" consists of Fontenelle's notions of the simple, tranquil purity of the shepherd's life. The virtues of the savant are not those of the nobility, but of the simple shepherd.

Fontenelle's pastoralism was not a random turn in the development of his moral philosophy but was a logical component of his steadfast reversal of Aristotelian formulae. For Aristotle, the concomitant to "man the rational animal" is "man the political animal." Men cannot perfect their nature as rational beings outside of the *polis*. For Fontenelle, however, man is a passionate animal, and he can therefore only perfect his nature if not corrupted by the *polis*—that is, by too much involvement in the mundane affairs of human society. His ambitions, immoderate desires, and passions had brought Fontenelle to despair, but the pastoral life, he wrote, "admits no ambition at all, nor those things which agitate the heart too violently; the idle [shepherd] therefore has cause to be content." This is because "the condition of the shepherd is the most ancient of all conditions" and filled with "tranquility and idleness."[66] For Fontenelle, shepherds are the essence of humanity, uncorrupted by the burdens of civilization: "They live in their own way in great opulence; they have no one above them, that is to say, they are the kings of their own flocks; and I do not doubt that a certain joy which follows such abundance and freedom inclines them to song and poetry."[67] Indeed, this is why "paintings of the pastoral life always have that inexpressible pleasantness, and [why] they soothe us more than the pompous descriptions of a superb court."[68] In order to attain happiness, goodness, and virtue, then, we must throw off the burdens of civilization and return mentally to a state of simple, Arcadian purity. Paradoxically, such a mental return to a mythic past was key to physical and technological progress in the very real future. As we see in Fontenelle's *éloges*, it is through contemplation, through natural philosophy, that he hoped humanity could achieve this "most delightful" state of tranquility.

Fontenelle's moral philosophy of the sciences laid the groundwork for a more radical and political vision of the savants' role in society in the eighteenth century—the beginnings of an Enlightenment political consciousness in which the idea of the natural sciences as morally enlightening demanded the active political leadership of savants. Moreover, his efforts to popularize the socially beneficial work of the sciences were so successful that philosophes and reformers of the eighteenth century felt they could at last call upon the authority of the sciences and the expertise of savants to justify their proposed reforms. As a moral thinker representing the work of the Academy of Sciences, Fontenelle stood as the mediator between the seventeenth and eighteenth centuries, between Sorbière and the philosophes who would take the moral idea of the sciences from its modest, humble beginnings as a supporter of social stability to the heights of the sublime as a means of reforming of the world.

NOTES

1. See, in particular, Anne Goldgar, *Impolite Learning: Conduct and Community in the Republic of Letters, 1680–1750*. See also Steven Shapin, *A Social History of Truth*.
2. See Smith, *Nobility Reimagined*.
3. Jonathan Israel also positions Fontenelle as a thinker highly critical of received ideas, religious in particular. See Israel, *Radical Enlightenment*, 359–74.
4. See Simone Mazauric, *Fontenelle et l'invention de l'histoire des sciences à l'aube des lumières* (2007).
5. Pierre Gassendi, *Peiresc, 1580–1637: Vie de l'illustre Nicolas-Claude Fabri de Peiresc, Conseiller au Parlement d'Aix*, translated by Roger Lasalle. On Peiresc and patronage, see Lisa Sarasohn, "Nicolas-Claude Fabri de Peiresc and the Patronage of the New Science in the Seventeenth Century," *Isis* 84 (1993).
6. See, specifically, Goldgar, *Impolite Learning*. Goldgar bases her thesis about French intellectual culture on Steven Shapin's theory about the seventeenth-century English scientific community in *A Social History of Truth*. The link between the "gentlemanly" and scientific ethic in England seems to have been stronger than in France, however. Alternately, at least in the particular case of Fontenelle, we need to qualify the application of Shapin's insights with a recognition of Fontenelle's critique of what he considered the failings of the noble honor ethic.
7. Roger Hahn argues against the decline and renewal of the Académie in *Anatomy of a Scientific Institution*.
8. See the éloge de Hombert, *Oeuvres Complètes de Fontenelle*, 1: 198. "L'académie, par le concours de quelques circonstances malheureuses, était tombée alors dans une assez grande langeur. Souvent on ne trouvait pas de quoi occuper les deux heures de séance." See also "Histoire du renouvellement de l'Académie Royale des Sciences en M.DC.XCIX," *Oeuvres Complètes de Fontenelle*, 1: 38–48.
9. Hahn, *Anatomy of a Scientific Institution*, 20; Sturdy, *Science and Social Status*, 214.
10. Hahn, *Anatomy of a Scientific Institution*, 20–34.
11. According to Alice Stroup's detailed analysis of the Academy, Colbert spent an average of 87,700 livres per year on the Academy, and Louvois only 26,484. Colbert allocated an average of 35,762 livres per year for pensions, and Louvois only 19,093. See Stroup, *A Company of Scientists*, 46–56.
12. Dates from Sturdy, *Science and Social Status*, 214–15.
13. Robert Briggs, "The Académie Royale des Sciences and the Pursuit of Utility," 48–56.
14. Stroup, *A Company of Scientists*, 56–57.
15. Sturdy, *Science and Social Status*, 221–22. Sturdy discusses the new regulations on 283–89.
16. This broke down to: ten full academicians, twenty *honoraires*, twenty *pensionnaires*, and twenty *associés* or *élèves*. "Réglement Ordonné par le Roi pour l'Académie royale des Sciences," in *Oeuvres Complètes de Fontenelle*, 1: 40–48.
17. See Birn, *Royal Censorship of Books in Eighteenth-Century France*, 9–20.

18. On Bignon's work for the Academy, and his relationship with Fontenelle, see J. A. Clarke, "Abbé Jean-Paul Bignon, 'Moderator of the Academies,' and Royal Librarian," 213–35.

19. In *The Fabrication of Louis XIV*, Peter Burke places the academies in the context of Louis XIV's efforts to express his *gloire*, and to reveal the harmony and order of his reign.

20. On the institution of the *éloges* in the Académie royale des sciences, see Charles Paul, *Science and Immortality: The* Éloges *of the Paris Academy of Sciences (1699–1791)*, 5–8. Also see Volker Kapp, "Les qualitiés scientifique et le prestige social des sciences dans les éloges académique de Fontenelle," in Alain Niderst, ed., *Fontenelle, actes du colloque tenu à Rouen du 6 au 10 octobre 1987*, 435–40.

21. On tropes in history, see Hayden White, *Tropics of Discourse: Essays in Cultural Criticism*.

22. See Alain Niderst, *Fontenelle*, 14–15, for a discussion of where and what Fontenelle studied.

23. Paul, *Science and Immortality*, 89–93.

24. See *Oeuvres Complètes de Fontenelle*, 3: 51–69, "Discours sur la nature de l'églogue." This will be discussed at greater length below.

25. *Oeuvres Complètes de Fontenelle*, 1: 62. "Il avait cette innocence et cette simplicité de moeurs que l'on conserve ordinairement, quand on a moins de commerce avec les hommes qu'avec les livres."

26. *Oeuvres Complètes de Fontenelle*, 1: 81. "Une simplicité, une franchise et une candeur, que le peu de commerce avec les hommes pouvait conserver."

27. *Oeuvres Complètes de Fontenelle*, 1: 95. "Les moeurs de Regis étaient telles que l'étude de la philosophie les peut former, quand elle ne trouve pas trop de résistance du côté de la nature."

28. *Oeuvres Complètes de Fontenelle*, 1: 132. "La vraie philosophie avait pénétré jusqu'à son coeur, et y avait établi cette délicieuse tranquillité, qui est le plus grand et le moins recherché de tous les biens."

29. *Oeuvres Complètes de Fontenelle*, 1: 201. "Une philosophie saine et paisible le disposait à recevoir sans trouble les différens événemens de la vie, et le rendait incapable de ces agitations dont on a, quand on veut, tant de sujets. A cette tranquillité d'ame tiennent nécessairement la probité et la droiture: on est hors du tumulte des passions."

30. *Oeuvres Complètes de Fontenelle*, 1: 402. "augmentée encore par la sage simplicité de sa vie."

31. *Oeuvres Complètes de Fontenelle*, 1: 337. "Son caractère était aussi simple que sa supériorité d'esprits pouvait le demander. J'ai déjà donné cette même louange à tant de personnes de cette académie, qu'on peut croire que le mérite en appartient plutôt à nos sciences qu'à non savans."

32. *Oeuvres Complètes de Fontenelle*, 1: 88–89. "Il faudrait maintenant le représenter comme homme, et peindre ses moeurs: mais ce serait le panégyrique d'un saint. . . . On voyait aisément que son humilité était, non pas un discours, mais un sentiment fondé sur sa science même."

33. On Neostoicism and the sciences in France, see the previous chapter. See also Robert Evans, *Jonson, Lipsius, and the Politics of Renaissance Stoicism*, and Margaret Osler, ed., *Atoms, Pneuma, and Tranquillity: Epicurean and Stoic Themes in European Thought.*

34. See Leonard Marsak, *Bernard de Fontenelle: The Idea of Science in the French Enlightenment.*

35. See Niderst, *Fontenelle*, 13–19.

36. *Oeuvres Complètes de Fontenelle*, 1: vi; Niderst, *Fontenelle*, 50–52. See also Grandjean de Fouchy, "Eloge de Fontenelle," *Histoire de l'Académie royale des sciences, de l'année MDCCLVII*, 282–306.

37. The late seventeenth century saw the beginning of a "rehabilitation" of the passions among thinkers who thought that Neostoics such as du Vair had gone too far in suggesting a total suppression of them. See Levi, *French Moralists*, and Albert O. Hirschman, *The Passions and the Interests: Political Arguments for the Capitalism before its Triumph.*

38. *Oeuvres Complètes de Fontenelle*, 2: 179. "Vous prétendez railler; mais je vous soutiens qu'il est plus difficile de boire et de chanter comme j'ai chanté et comme j'ai bu, que de philosopher comme vous avez philosophé. Pour chanter et pour boire comme moi, il faudrait avoir dégagé son âme des passions violentes, n'aspirer plus à ce qui ne dépend pas de nous, s'être disposé à prendre toujours le temps comme il viendrait."

39. *Oeuvres Complètes de Fontenelle,* 2: 215. "Ce sont les passions qui font et qui défont tout. Si la raison dominait sur la terre, il ne s'y passerait rien. On dit que les pilotes craignent au dernier point ces mers pacifiques où l'on ne peut naviguer, et qu'ils veulent du vent, au hasard d'avoir des tempêtes. Les passions sont chez les hommes des vents qui sont nécessaires pour mettre tout en mouvement, quoiqu'ils causent souvent des orages."

40. *Oeuvres Complètes de Fontenelle*, 2: 224. "Elle a mis les hommes au monde pour y vivre; et vivre, c'est ne savoir ce que l'on fait la plupart du temps. Quand nous découvrons le peu d'importance de ce qui nous occupe et de ce qui nous touche, nous arrachons à la nature son secret: on devient trop sage, et on ne veut plus agir; voilà ce que la nature ne trouve pas bon."

41. *Oeuvres Complètes de Fontenelle*, 2: 242–43. "Toutes les grandes actions qui doivent être faites pas les hommes se trouvent faites: enfin, l'ordre que la nature a voulu établir dans l'univers va toujours son train." Marsak's translation. See Leonard Marsak, *Bernard de Fontenelle*, 9.

42. *Oeuvres Complètes de Fontenelle*, 2: 190–91. "Les hommes de tous les siècles ont les mêmes penchans, sur lesquels la raison n'a aucun pouvoir. Ainsi, partout où il y a des hommes, il y a des sottises, et les mêmes sottises."

43. *Oeuvres Complètes de Fontenelle*, 2: 246. "J'assemblais dans un certain lieu le plus grand nombre de gens que je pouvais, et là je leur faisais voir qu'ils étaient tous des sots. . . . Rien n'est plus facile. On leur prouve leurs sottises, sans employer de grands tours d'éloquence, ni des raisonnemens bien médités. Ce qu'ils font est si ridicule, qu'il ne faut qu'en faire autant devant eux, et vous les voyez aussitôt crever de rire."

44. *Oeuvres Complètes de Fontenelle*, 2: 247. "Pour rire des choses du monde, il faut en quelque façon en être dehors, et la comédie vous en tire: elle vous donne tout en spectacle, comme si vous n'y aviez point de part."

45. Leonard Marsak, *Bernard de Fontenelle*, 8–12, offers a somewhat more positive interpretation.

46. See Niderst, *Fontenelle à la Recherche de lui-même*.

47. Fontenelle, *Conversations on the Plurality of Worlds*, translated by H. A. Hargreaves, 7, 11.

48. For a few discussions in English, see Geoffrey Sutton, *Science for a Polite Society*, especially chapter 5, "A Pretty Novel of Physics, in which *cogito, ergo sum* meets *l'état c'est moi*," 143–87; see also S. J. Dick, *Plurality of Worlds: The Origins of the Extraterrestrial Life Debate from Democritus to Kant*.

49. There are no clear indications I have seen of when Fontenelle wrote the essay. It was only published in 1724 in a collection of diverse works all written more than twenty years before. I suggest he wrote it in the 1680s, or at most the early 1690s, based on its congruence with several other works from the late 1680s. The issues the essay raises, and especially the language it uses, link it to a 1687 essay, "Discours sur la patience" (for which Fontenelle won the Académie Française's prize for eloquence), as well as to his 1687 essay on the nature of the Eclogue (a poem in which shepherds converse). The appearance of "Du Bonheur" in 1724 along with other early essays seems a rather self-indulgent afterthought—perhaps the act of a famous and elderly writer (he was in his late sixties) publishing his works for the benefit of his public.

50. *Oeuvres Complètes de Fontenelle*, 2: 370. "Quand ce Stoïcien, pressé par la douleur d'une maladie violente, s'écriait, en s'adressant à elle: Je n'avouerai pourtant pas que tu sois un mal; cet effort qu'il faisait pour ne le pas avouer, ce désaveu même apparent, n'était-ce pas un aveu et le plus fort et le plus sincère qui pût jamais être?"

51. *Oeuvres Complètes de Fontenelle,* 2: 378. "On entend ici par le mot de bonheur un état, une situation telle qu'on en désirât la durée sans changement."

52. *Oeuvres Complètes de Fontenelle*, 2: 379. "La ridicule et inutile vanité de nous croire invulnérables."

53. *Oeuvres Complètes de Fontenelle*, 2: 380. "Il en reste quelques-unes, doux et modérés, et qui admettent plus volontiers les idées ou les impressions agréables: ceux-là peuvent travailler utilement à se rendre heureux."

54. *Oeuvres Complètes de Fontenelle*, 2: 380. "Chassé tous les maux imaginaires."

55. *Oeuvres Complètes de Fontenelle*, 2: 384. "On ne daigne pas s'arrêter à goûter ceux que l'on possède; souvent on les abandonne pour courir après ceux que l'on n'a pas."

56. *Oeuvres Complètes de Fontenelle*, 2: 385, 386. "Pourquoi cette dignité que je poursuis m'est-elle si nécessaire? c'est qu'il faut être élevé au-dessus des autres. Et pourquoi le faut-il? c'est pour recevoir leurs respects et les hommages. Et que me feront ces hommages et ces respects? Ils me flatteront très-sensiblement. Et comment me flatteront-ils, puisque je ne les devrai qu'à ma dignité, et non pas à moi-même? . . . Souvent le bonheur dont on se fait l'idée, est trop composé et trop compliqué. Combien de choses, par exemple, seraient nécessaires pour celui d'un courtisan? du crédit auprès ministres, la faveur du Roi, des établissemens considérables pour lui et pour ses enfans, de la fortune au jeu, des maîtresses fidèles et qui flatassent sa vanité; enfin tout ce que peut lui représenter une imagination effrénée et insatiable."

57. *Oeuvres Complètes de Fontenelle*, 2: 385–86. "Il en faut revenir aux plaisirs simples, tels que la tranquillité de la vie, la société, la chasse, la lecture, etc. . . . Les gens accoutumés aux mouvemens violens de passions, trouveront sans doute fort insipide tout le bonheur que peuvent produire les plaisirs simples. Ce qu'ils appellent insipidité, je l'appelle tranquillité. . . . Mais quelle idée a-t-on de la condition humaine, quand on se plaint de n'être que tranquille? et l'état le plus délicieux que l'on puisse imaginer."

58. See Keohane, *Philosophy and the State in France*, 83–116. See also Charles Taylor, *The Sources of the Self: The Making of the Modern Identity*, 177–84. Compare Fontenelle's critique of the nobility as immoderate to Montaigne's in his essay "On Moderation," *The Complete Essays*, translated by M. A. Screech, 222–23: "I have seen one of our great noblemen harm the reputation of his religion by showing himself religious beyond any example of men of his rank. . . . I like natures which are temperate and moderate. Even when an immoderate zeal for the good does not offend me, it still stuns me."

59. *Oeuvres Complètes de Fontenelle*, 2: 387. "Le plus grand secret pour le bonheur, c'est d'être bien avec soi. Naturellement tous les accidens fâcheux qui viennent du dehors, nous rejettent vers nous-mêmes, et il est bon d'y avoir une retraite agréables; mais elle ne peut l'être si elle n'a été préparée par les mains de la vertu."

60. *Oeuvres Complètes de Fontenelle*, 2: 387 "Mais une récompense infaillible pour elle, c'est la satisfaction intérieure . . . on trouve dans sa propre raison et dans sa droiture un plus grand fonds de bonheur que les autres n'en attendent des caprices du hasard."

61. Jonathan Barnes, "Introduction" in the *Ethics of Aristotle*, translated by J. A. K. Thomson, 36.

62. *Oeuvres Complètes de Fontenelle*, 2: 399–406. Fontenelle's understanding of reason has been considered in several studies, notably Marsak, *Bernard de Fontenelle*; J.-R. Carré, *La Philosophie de Fontenelle, ou le sourire de la raison*; P. Flourens, *Fontenelle, ou de la philosophie moderne relativement aux sciences physiques*; and F. Grégoire, *Fontenelle: Une philosophie désabusée*.

63. *Oeuvres Complètes de Fontenelle*, 1, 201. "Quiconque a le loisir de penser, ne voit rien de mieux à faire que d'être vertueux."

64. See especially his "Digression sur les anciens et les modernes," *Oeuvres Complètes de Fontenelle*, 2: 353–55, which asserts that since human nature is constant, there is no moral difference between the ancients and moderns. Since scientific knowledge is cumulative, however, the moderns are more advanced in this area than the ancients.

65. See Jay Smith, *Nobility Reimagined*.

66. *Oeuvres Complètes de Fontenelle*, 3: 57. "Elle n'admet point l'ambition, ni tout ce qui agite le coeur trop violemment; la paresse a donc lieu d'être contente." Ibid., 3: 52: "la condition de berger est la plus ancienne de toutes les conditions."

67. *Oeuvres Complètes de Fontenelle*, 3: 52. "Ils vivaient à leur manière dans une grande opulence, ils n'avaient personne au-dessus de leur tête, ils étaient pour ainsi dire les rois de

leurs troupeaux; et je ne doute pas qu'une certaine joie qui suit l'abondance et la liberté, ne les portât encore au chant et à la poésie."

68. *Oeuvres Complètes de Fontenelle,* 3: 57. "Les peintures de la vie pastorale aient toujours je ne sais quoi de si riant, et qu'elles nous flattent plus que de pompeuses descriptions d'une cour superbe."

Chapter Three

Fear and Loathing in the Courts of Louis XV and Louis XVI: The Sciences and the Crisis of the Monarchy from Voltaire to Turgot

When Bernard de Fontenelle retired from his post as *secrétaire perpétuel* of the Royal Academy of Sciences in 1740, after some forty years of service, the authority and prestige of the Academy, of the academicians, and of the natural sciences in general had risen almost immeasurably in France as well as in the rest of Europe. Fontenelle cannot take all the credit for the enhanced prestige of the sciences, of course. It was largely through the efforts of Bignon, whose active and enlightened administration breathed new life into the Academy, that Fontenelle's work was made possible. Important achievements by academicians such as Tournefort (his system of plant classification, outlined in the *Institutiones rei herbariae*, 1700 and 1719) and Homberg (his *Essai de chimie*, 1702–1709) also raised the profile of the natural sciences considerably, as did several scientific voyages to the Levant (1700), Peru (1735), and Lapland (1736)—all of which gave Fontenelle noteworthy subjects about which to write. Nevertheless, among those savants and philosophes—generally termed the *parti philosophique*—who saw the natural sciences as a source of moral enlightenment that promoted civilized values and the progressive reform and amelioration of society, Fontenelle was held in high esteem for his role in popularizing a new culture of the sciences, a *goût des sciences*, in France and Europe among the educated public. After all, scientific achievements did not always happen quickly or dramatically, and much work was carried on away from the public eye or could not be understood unmediated except by experts. It was, therefore, necessary to put the sciences in the proper philosophical context. The need to explain and instruct the public on the benefits of the sciences formed a large part of Bignon's mission when he took over the Academy in 1691. To accomplish this mission, he opened Academy meetings to the public, arranged to have the *Histoires et mémoires* of the Academy published annually, and took over the ailing *Journal des sçavans* as an official Academy publication.

Fontenelle served as a primary instrument of this mission and did much of the quotidian labor. The detailed biographical eulogies and histories Fontenelle composed and presented publicly also served to outline and explain each academician's work to the nonspecialist and to highlight systematically the moral excellence of the scientific enterprise. His work tapped into the ethical debates of the late seventeenth and early eighteenth centuries, and it positioned savants as *ésprits supérieurs*, a natural moral elite standing out from a crowd of weak and mediocre minds and engaged in universally beneficial labor. The effect of Fontenelle's writing was to clarify and popularize a view of the sciences that implied a special role for savants in the establishment of social tranquility, progress, and civilized values (see chapter 2). Fontenelle made no overt suggestions of savants' political role, but what was implicit in the moral philosophy of the sciences Fontenelle articulated was made explicit by later science moralists, particularly by the marquis de Condorcet in the 1770s. Savants and philosophes, enjoying enhanced prestige and popularity, began to take seriously the idea that they had a political role to play. Until the end of the eighteenth century, however, core values changed little: those who promoted scientific enlightenment remained bound to the basic pessimism of their Neostoic philosophical roots. Philosophes such as Voltaire continued to argue, as had Sorbière and Fontenelle, that the desired social, economic, and political change in France must come from above and that the goal of the new moral elite was to help guide this change through a rationally organized central administration that governed society according to scientific principles. Even when Condorcet, among others, later decided that universal education in the sciences and critical reasoning could perform an emancipatory function, he continued to consider that a natural moral and intellectual elite would emerge from society to direct and protect its progress. Perhaps the biggest difference between pre- and post-Fontenellian *philosophie* was a decline in intellectual modesty. As we shall see, the faction of the *parti philosophique* closely associated with Voltaire—which, as Jonathan Israel has recounted, experienced a break with the philosophical party as a whole starting in the late 1730s—succumbed to an almost astonishing degree of hubris.[1]

Savants and philosophes had reason for optimism, of course, hopeful that the state would continue to support the progress of the sciences and that the sciences could assist in what many of them considered the necessary rationalization and reform of society. And yet the era from 1740 through 1776 was one of almost desperate striving that finally gave way to dark despair—particularly among those ideologically associated most closely with Voltaire. The apotheosis of the Voltairean faction's hopes was the rise of their associate, the philosophe and economist Robert-Jacques Turgot, to the king's ministry in 1774. Turgot sought to implement profound reform policies intended to transform the political, economic, and even social structure of Old Regime France to correspond to the rationalizing ideals of the philosophes. Perhaps first among the great ministers of state, Turgot tried to legitimize unpopular reforms with the authority of scientific reason, and to raise savants—and the Royal Academy of Sciences—to an active participatory role in state functions. In essence, Turgot's ministry was intended to be the culmination of a century-long dream in which savants finally stood at the helm of state with power and reason joined in perfect union. As a result, Voltaire's associates interpreted Turgot's spectacular failure and fall from power in 1776 as a repudiation of their ideals, as the opening of a vast chasm between their desired reforms and the means believed hitherto necessary for achieving them. The moral idea of the sciences had always gone hand in hand with the ideal of government power and authority. The necessity to deal intellectually with the failure to unify power and reason paved the way for a radicalization of political goals, especially with the youngest savant of Voltaire's faction, the marquis de Condorcet (whose ideological transformation I will explore in chapter 4).

There are various ways one might measure or roughly gauge the diffusion of the new culture of the sciences in the eighteenth century. Historians, for example, have explored the fashion for scientific experimentation and scientific speculation in the Parisian salons, where they became the subjects of discussion and were even used as objects of entertainment.[2] The sciences were not merely a salon game, however, and we might also comprehend the influence of the Royal Academy of Sciences of Paris by the number of imitators it had. In the provinces, some fifteen academies arose between 1700 and 1750, especially in major cities. Montpellier formed its own *Société royale des sciences* in 1706, and Bordeaux the *Académie royale des sciences et belles-lettres* in 1712. Lyon followed suit in 1724, Dijon in 1725, and Brest in 1752. In 1744 Fontenelle himself helped secure royal patents for a science society in his native city of Rouen. Beyond France, European rulers were eager to appear as enlightened as the French kings and to secure for themselves the benefits of state-sponsored scientific and technological innovation—as well as the new political legitimacy the sciences seemed to promise as a source of rational authority. Under Leibniz's guidance, Frederick I of Prussia formed the *Societas Regia Scientiarum* in 1700 (reformed in 1746 under the Francophile Frederick II as the *Académie royale des sciences et belles-lettres de Prusse*). After a visit to the Academy in Paris in 1717, during which he was made an honorary member, Czar Peter the Great of Russia spoke with Fontenelle and other savants on the possibility of forming a science academy in St. Petersburg. His schemes came to fruition in 1724 as the Imperial Russian Academy of Sciences, an institution staffed largely by well-paid foreign savants. At roughly the same time, formal academies opened in Uppsala, Stockholm, and Bologna.[3]

The fact that not only the cultural and intellectual elite but also European rulers in the eighteenth century were coming to see both savants and the natural sciences as important servants of the state is obviously significant for our understanding of the development of the Enlightenment. The desire to establish social and political order—even discipline—after the political crises of the sixteenth and seventeenth centuries strongly motivated proponents of strong, centralized state authority to support an absolute (and absolutist) monarchy (see chapter 1).[4] Many of the advocates of absolutist monarchy—the intellectual descendants of the *politiques*, Neostoics, and, of course, *libertins érudits*—believed the sciences offered the sort of rational, practical, and nondogmatic authority that could best serve the state and help to establish its legitimacy on a firm basis, at least in the eyes of the politically significant public (little thought was given to the people as a whole). The push for state-sponsored scientific academies after the Frondes was therefore part of an effort to stabilize the monarchy by transforming it from a patrimonial (perceived as arbitrary) to a bureaucratic and even a sort of technocratic (perceived as rational) administrative apparatus.[5] In the first two-thirds of the eighteenth century in France, the high point for royal academies of the arts and sciences, the ideological connection between absolute monarchy and the sciences, continued to dominate the consciousness of proponents of centralized discipline. We commonly term their political ideology *enlightened despotism*, but philosophes tended not to speak of their ideal government as despotic in the negative sense, but rather as legally or constitutionally absolute in that it based undivided political power on rational laws discovered through natural and political sciences, submitted to the review of a small but enlightened "public."[6] Although Peter Gay once argued that the "science of freedom" was a primary goal of the Enlightenment, I would suggest that, at least until late in the eighteenth century, the more pressing goal was a "science of order" and, perhaps, of justice.[7] Liberty, in its pure form, was a much more problematic issue; in the hands of nobility and parlements "traditional liberties" could be perceived as a source of corruption and injustice—since they were, strictly speaking, neither traditional nor liberties but rather privileges. Only an enlightened and absolute government in France, many philosophes argued, could protect the people from mistreatment at the hands of autonomous privileged classes such as the nobility, the parlements, and the Church, whose interests were not those of the nation as a whole even if they pretended they were. What was needed was justice, not privileges disguised as liberties—an absolute and unassailable justice, ensured by order and stability that prevented the exercise of arbitrary and unaccountable power. An absolutist monarch supported and advised by a ministry of enlightened men and by institutions of scientific research seemed to offer a solution to the problem of establishing order and justice. In short, the sciences and the state had ideological as well as instrumental reasons for supporting each other. In fact, it would be impossible to separate the ideological reasons from the instrumental ones.[8]

It is easy to see why Charles Gillispie, in his comprehensive study of science and the state at the end of the Old Regime, did not view the sciences as a source of a liberal or progressive reform movement.[9] The sciences, or rather a moral-political discourse based on an idea of the sciences as morally enlightening, were an intricate part of an ideology that positioned absolute monarchy as a political ideal. The tendency of historians to pose an almost Hegelian dialectical opposition between absolutist monarchy and the progressive and reform-minded political thinkers that made up the supposedly egalitarian public sphere of the Republic of Letters is anachronistic.[10] Such a view also tacitly accepts the propaganda of noble moralists, such as Boulainvilliers, Fénelon, and Montesquieu, who criticized the monarchy as despotic in order to promote the role of the nobility as the basic check to royal despotism. Many philosophes, however, saw enlightened absolute monarchy as the guarantor of necessary and progressive reforms—as Voltaire said to d'Alembert in 1765, "The cause of the king is the cause of the philosophes."[11] The royal government, particularly during the short ministry of Robert-Jacques Turgot, did not seek in the sciences merely powers to be used as it saw fit, but instead sought a merging of political and rational scientific authority that would raise the government's legitimacy to an unquestionable level—such that resistance to the central administration would be tantamount to denying the dictates of truth and reason. Keith Baker concluded that the collaboration of government and scientific institutions during the ministry of Turgot was a response to what he calls a "crisis of legitimation that afflicted the government."[12] Beyond mere practical collaboration, however, the era of Turgot witnessed a hitherto unprecedented attempt to merge the state with the rational sciences in order to resolve the long-standing crisis of social and political order—a crisis that the French in the 1760s experienced all too well with popular riots, political scandals, the Brittany Affair, and the commotions of the Maupeou coup.[13] By raising the moral, practical, and political authority of the monarchy beyond all rational appeal, proponents of an enlightened absolute monarchy sought to either sway the opinions or force the acquiescence of their opponents in the face of popular support for a rational, effective, and just administration.

The move to establish what we might now term a proto-technocracy under Turgot should not be viewed as a sudden infatuation of the state with the political possibilities of the sciences. Turgot's ministry was the culmination of an ideal of enlightened administration that began in the seventeenth century with Sorbière and the founders of the Royal Academy of Sciences. The dramatic failure of Turgot and his allies to accomplish the merging of state and sciences led to a moment of despair that sparked an interesting transformation in the idea of the sciences as morally enlightening: a move to use the sciences to emancipate the people through education, and thus to accomplish reforms from below, as it were, rather than from the top down.

One thing is clear enough: By the middle of the eighteenth century, the sciences had *arrived* socially among the sophisticated cultural elite of the Parisian salons and the Republic of Letters. To paraphrase Fontenelle, the sciences were like a beautiful young marquise making her debut in society—a new morsel to be pursued for the amusement of the fashionable and urbane. Despite this, savants and philosophes still had to insist, often quite shrilly, on the social and political utility of the sciences. In *Le Siècle de Louis XIV* (The Century of Louis XIV—begun in 1733, and published in 1751), Voltaire noted—with an ironic satisfaction that rhetorically disguised his nagging doubts about prevalent Old Regime mores—that the natural sciences had brought about a "revolution in the human spirit" that was rapidly ridding the world of superstition, ignorance, and error.[14] Likewise, d'Alembert, in the "Discours préliminaire" of the *Encyclopédie* (1751), observed with a faux confidence that the philosophic spirit, the dominant fashion of his century, seemed to be making up for time lost in the Dark Ages.[15] By the 1750s, such lines were familiar—little more than paraphrases of Fontenelle—but we should not take them at face value to indicate that the sciences had achieved victory.[16] To the *parti philosophique*—and especially the Voltairean faction—there was, nevertheless, room for hope.

Indeed, the rising tide of state support for scientific work instilled a new confidence in the scientific community and in the Royal Academy of Sciences in particular. This confidence is revealed in the choice of Fontenelle's successors as *secrétaires perpétuels*. Bignon had brought Fontenelle into the academy as a writer and philosopher, not as an expert in any particular science, to perform a special function as a popularizer, a marketer, at a critical moment in the history of the Academy. Apparently no longer sensing a need for such a personality as their public face, the Academy chose his successors from the ranks of the Academy itself: first, the mathematician Jean-Jacques d'Ortous de Mairan, secretary from 1741 to 1743, and then the astronomer Jean-Paul Grandjean de Fouchy, secretary from 1743 to 1773. The difficulty, even impossibility, of finding a replacement of Fontenelle's caliber led to the temporary appointment of Mairan, nearing retirement age himself, as interim secretary.[17] Eventually the Academy decided it was no longer necessary to replace Fontenelle with a philosopher of similar stature and appointed the thirty-five-year-old astronomer, Fouchy, merely to fulfill the secretary's duties as Fontenelle had more or less defined them. In the eyes of the academicians, Fontenelle had made himself obsolete.

In 1788, Fouchy's successor, Condorcet, provided his own interpretation of the Academy's new outlook at the time of Fontenelle's retirement—although, again, Condorcet's words should not be taken as unproblematic statements of fact, any more than those of Voltaire. In his eulogy of Fouchy, Condorcet noted that Fontenelle had expressed a *philosophie profonde* at the exact moment when "the natural march forward of the human spirit had marked for the people of Europe the passage from slavery to prejudices to the reign of reason." By 1740, however, Fontenelle's brilliant subtlety was no longer necessary, and the secretary could focus on performing his duties with simplicity, truth, and exactitude. As Condorcet remarked, "One [Fontenelle] reminded us of the general utility of the sciences, the beauty of the spectacle that they present to the human spirit, and the noble exercise that they offer to its activity; the other [Fouchy], speaking to men already impassioned by the sciences, had to limit himself to making clear the utility of each project for the progress of the science to which it pertained. The histories of M. de Fontenelle had to be more philosophical, those of M. de Fouchy more learned."[18]

Fouchy himself felt the weight of Fontenelle's shade upon the office he now occupied, leading him to ruminate on the duty of the perpetual secretary and the public role of the sciences. In his eulogy of Fontenelle in 1757, he reminded his auditors of Fontenelle's seemingly prophetic words in the preface to the *Histoire de l'Académie*: "Sometimes a great man sets the tone for his century," Fontenelle had written. "He himself was this great man that he announced," proclaimed Fouchy, "and we can regard him as one of those to whom the Sciences, and therefore all men, are the most obligated, and as a model that those who succeed him ought always to follow."[19] Succeeding the man whom many credited with helping to drive away the *espèce de barbarie* of previous eras and of communicating to the educated public the new *goût des sciences*, however, was daunting.[20] Most of Fouchy's duties as secretary had been defined and perfected by Fontenelle, especially the business of eulogizing. When it came time to publish his own eulogies, Fouchy could not do so without some sort of apology.[21] Nevertheless, Fontenelle had made clear the importance of the academic eulogy, and Fouchy admitted he knew of no better way to instruct youths drawn to the sciences than by historical and biographical eulogies of academicians. It was therefore the duty of the secretary to continue to write, deliver, and publish his eulogies and histories, even if his literary skill could not match Fontenelle's. His colleague, d'Alembert, agreed: "If the ancients who raised statues of great men had had the same care as we to write the lives of men of letters, we would have, it is true, some useless memoirs, but we would be more instructed on the progress of the arts and sciences, and on the discoveries of every age, a history more interesting to us than those of a crowd of sovereigns who have only done evil toward men."[22]

In elucidating a moral philosophy of the sciences, Fontenelle had in effect justified the role savants had begun to assert for themselves since the time of Mersenne and Sorbière. For philosophes such as d'Alembert, this role took on clear political overtones. Since progress in the arts and sciences was a universal good, savants had a public duty to educate, instruct, and guide the public. Such authority could not be exercised effectively, however, unless the public reputation of the learned academies for expertise, moral probity, and rational objectivity was assured. Fortunately, the Royal Academy of Sciences owed its public reputation to Fontenelle, d'Alembert asserted: "Without the art with which this celebrated writer popularized the works of his colleagues, these works, although excellent, would be known only to savants; they would remain unknown by what we call the *public*, and the reputation that the Academy of Sciences enjoys would be less universal."

This did not mean, d'Alembert continued, that Fontenelle's style was to be copied: "His style of writing pertained to him alone, and cannot be transferred, without losing it, to another pen; it is a liquor that ought not be rebottled."[23] In the new age, one must take a new tone, a new method, that assumed rather than asserted the public role of the sciences. Condorcet would announce in 1783 what Sorbière had only dreamed of 120 years earlier. "There is no more need to tell princes that they have an interest in protecting the sciences," wrote Condorcet, "or the public that the sciences have a right to its gratitude."[24] In fact, he voiced an attitude that had already gained considerable currency among philosophes by the middle of the century. Voltaire, too, pretended a supremely confident sensibility about the sciences in his philosophical story, *Micromégas* (written in 1739 but published in 1752), while at the same time poking good-natured fun at Fontenelle for what he suggested was a then outdated and precious style.[25] When the secretary of the Science Academy of Saturn, a thinly veiled caricature of Fontenelle, spoke in baroque terms of nature, comparing it to flowers, beautiful women, and galleries of paintings, the cosmic traveler, Micromégas, demanded impatiently the reason for all the comparisons. "To please you," answered the secretary, to which Micromégas retorted, "I do not wish to be pleased; I wish to be instructed."[26] Learned opinion held that it was no longer necessary, or even properly speaking in good taste, for the scientific community to seek to please the powers that be, but rather to instruct, guide, advise, and serve their rational interests. The sciences, the *parti philosophique* insisted, had arrived. But saying it ad nauseam did not make it so.

The true successor of Fontenelle as the laureate of science popularization in the mid-eighteenth century was, of course, Voltaire, not the *secrétaires per-*

pétuels of the Academy. Despite the pose of supreme confidence, the sciences needed defenders, and the idea of the sciences as morally enlightening needed promoters. Voltaire may not have officially succeeded Fontenelle, but after Fontenelle the sciences had no more eloquent a spokesman or defender than Voltaire, even if not all savants agreed with his views of the sciences. Serving as the intellectual link between the latter *libertins érudits*, such as Fontenelle and even Pierre Bayle, and the younger generation of philosophes, Voltaire was crucial in advancing the moral and political values that had arisen in and around the Academy, and which had hitherto been so well articulated by Fontenelle.

As early as the 1720s Voltaire, following on Fontenelle's success, had set himself out to become a master popular science writer.[27] His first serious accomplishments were the *Lettres philosophique* (1733) and the *Eléments de la philosophie de Newton* (1736). Perhaps out of vanity, Voltaire rarely quoted or cited Fontenelle except to criticize his unfortunate defense of Cartesian vortices against the Newtonian theory of attraction. Voltaire's expertise in the modern sciences never extended much further than Fontenelle's, however, whose eulogy of Newton served as the basis for Voltaire's own Newtonian writings.[28] The very idea of science popularization undeniably followed Fontenelle's lead, of course, and Voltaire also adopted Fontenelle's goal, if not his style. For both, the purpose of popular science writing was not to demonstrate technical expertise in any particular science but to distinguish clearly and boldly the right kind of science from the wrong kind—and even the right kind of thinking from the wrong kind.[29] Here Voltaire and Fontenelle were, more or less, of one mind: the right kind of science was experimental and practical, and it avoided gauche expressions of systematic or dogmatic inflexibility.

Voltaire's engagement in science popularization was not merely part of a vain attempt to become the undisputed master of all literary genres in France. Voltaire was more than the literary scion of the *libertins érudits*; he agreed with them politically and philosophically. Early in his career, Voltaire identified clearly with *politique* ideology and remained a defender of enlightened absolute monarchy in France—although not necessarily elsewhere—until the end of his life. Like the *politiques*, Neostoics, and early *libertins érudits*, he even had an example of undignified aristocratic intransigence and petty self-interest against which to respond: the renewal of aristocratic resistance to the absolute monarchy during the regency, immediately after the death of Louis XIV in 1715. By 1716, and perhaps as early as 1715, Voltaire commenced his epic poem *La Henriade* (published in 1723) in response to aristocratic maneuvering. *La Henriade* painted King Henry IV, the founder of the Bourbon dynasty, as an idealized hero and *politique* who defeated the evil Catholic League and the seditious factions of the kingdom, and who restored tranquility to France by strengthening the monarchy. Moved by tolerant humanitarian principles rather than dogmatic religious ones, he also sought to improve the lot of the French people in this world, not to save souls or protect the interests of the Church and aristocracy.

La Henriade marked Voltaire as a *politique* himself, but the poem was only the beginning of his combat against noble pretensions—a battle that became even more acerbic after his humiliating beating at the hands of the Chevalier de Rohan's thugs in 1726, followed by a short stint in the Bastille. *La Ligue* (1723), reviling the infamous Catholic League, was anti-noble propaganda, as were his first historical and philosophical works. The *Histoire de Charles XII* (1731), written while in exile in England, examined the struggle for political authority and public order in Sweden between monarchy and nobility (the latter allied with the Church). Asserting that this struggle existed in virtually every European country, Voltaire concluded that the nobility, in seeking to retain its feudal privileges, was everywhere the enemy of modernity and tranquility.[30] The *Lettres philosophiques* made similar caustic comments: parliaments (and parlements?) were in origin assemblies of "ecclesiastical tyrants and robbers called barons" under which the people groaned in miserable thralldom. The great blessing of England was the weakening of the House of Lords beginning with Henry VII, a development that brought equal justice and rational taxation to the kingdom. "Is it not fortunate for the human race," he wrote sarcastically, "that the authority of these petty brigands was destroyed in France by the legitimate power of kings and people?"[31]

Immediately upon the publication of the *Lettres philosophiques*, Voltaire began amassing documents for his most ambitious historical treatise, *Le Siècle de Louis XIV*, which would not appear in even its earliest forms until the late 1740s and saw its first printing in 1751. In the meantime, he completed various other essays, including the *Essai sur l'histoire générale*, which eventually evolved into the famous *Essai sur les moeurs* (1756). The *Essai*, influenced by the comte d'Argenson's *Considérations*, which Voltaire read in manuscript form in 1739, and the equally royalist abbé Dubos's *Histoire critique de l'établissement de la monarchie française* (1734), decried feudal aristocratic privileges as sources of anarchy that were detrimental to art, science, and civilized values.[32] The same judgment appeared again in *Le Siècle de Louis XIV*. Although critical of the later years of Louis XIV's reign, particularly of Louis's wars of aggression in the 1670s and 1680s and his disastrous decision to renew Protestant persecution, *Le Siècle de Louis XIV* was Voltaire's most detailed and exhaustive defense of absolute monarchy to date, lauding what he argued was one of the four great epochs in the history of civilization and "the most enlightened century there ever was." The main reason for the greatness of the era was the flourishing of a *saine philosophie* and its civilized virtues—and the support savants and philosophes gave to the monarchy's attempt to establish political order and social stability: "No one would believe that sovereigns were obligated to philosophers," wrote Voltaire. "It is true, however, that this philosophical spirit, which has penetrated to nearly all classes, except the lowest, has contributed much to legitimize the rights of sovereigns. The quarrels that produced in other times excommunications, interdictions, and schisms, caused them no longer. If it has been said that the people would be happy when they had philosophers for kings, it is also true to say that kings are even happier when many of their subjects are philosophers."[33]

The age of Louis XV, on the other hand, was one of growing political crisis to which Voltaire, in frustration and despair, lent an infectious apocalyptic perspective. The nearly finished *Siècle de Louis XIV* was not eagerly awaited at Versailles, where Voltaire was then royal historiographer: the book's high praise of Louis XIV implied criticism of his less able successor. The monarch was right to feel insecure about his abilities. Louis XV did not have his predecessor's charisma or personality, and he often failed to execute vigorously what Voltaire saw as the best interests of the monarchy over those of the Church and parlements. The popularity of noble resistance to "royal despotism" was high, especially after the appearance of Montesquieu's *De l'esprit des lois* in 1748, which raised the arguments for noble resistance to their highest philosophical level. Supporters of absolute monarchy found themselves taking a defensive stance. The Assembly of the Clergy, which began to convene in late 1749 and finally opened in June 1750, revealed the weakness of Louis XV in vivid relief. Despite intensive negotiations, the clergy refused to surrender its tax privileges or submit to secular authority, and Louis XV was too timid, superstitious, or conservative to press the issue.[34] Voltaire pressed it for him. In various pamphlets, most notoriously in *La voix du sage et du peuple* (1750), Voltaire viciously attacked the esprit de corps of the clergy and raised his voice in defense of undivided political authority: "The prince must be absolute master of all ecclesiastical regulations, without any restriction, since those ecclesiastical regulations are part of the government."

In *La voix du sage*, Voltaire also took care to insist that, while priests and clergy had never failed to be the king's unyielding opponents, his best and most sure friends were always philosophes: "There is in this world not a single example of philosophers who have opposed the laws of the prince; there is not a single century in which superstition and enthusiasm have not caused troubles which horrify us." In reminding the king of the natural alliance between monarchs and the disciplined defenders of reason, science, and modernity, Voltaire made some of his clearest statements yet for enlightened absolute monarchy: "The best thing that can happen to men is for the prince to be a philosopher. The philosophic prince knows that the more progress reason makes in his state, the less damage will be done by disputes, theological quarrels, enthusiasm, superstition; hence he will encourage the progress of reason."[35]

La voix du sage, like its author, was too audaciously anti-clerical for its time, however. Voltaire appeared to be more in favor of absolute monarchy than the king. For his own protection, or so he apparently thought, Voltaire accepted the invitation of his perhaps enlightened but definitely despotic admirer, Frederick II of Prussia, to join him at his court in Potsdam. Despite his exile and his misadventures with the cynical, tendentious ruler of Prussia, Voltaire continued to defend enlightened absolute monarchy in France—not for theoretical or doctrinaire reasons, but as the most practical means of effecting discipline, order, and reform. Even the gradual amelioration of his elitist attitude toward the *esprits faibles* after escaping Frederick and establishing his abode in Geneva in 1754 did not change his basic political views. While Voltaire defended the "democratic" interests of the bourgeoisie of Geneva, he continued to attack the parlements of France for resisting the authority of the monarchy. The alternatives were, to Voltaire, unrealistic or dangerous: despite their claims of popular representation and natural rights, the parlements had consistently shown themselves, in Voltaire's opinion, to be greedy, reactionary, intolerant, and hypocritical. What other alternatives were there? After 1762, there were Rousseau's suggestions as outlined in the *Social Contract*. But Voltaire found Rousseau's particular republican notions abhorrent, especially his intolerant and dogmatic ideas about a "civil religion."[36] Such notions went against the grain of Voltaire's central political philosophy: that social tranquility and political order rest on a just government of rational laws and scientific ideals. The goal of political leaders should not be to inculcate superstition and error but to dispel them by spreading the philosophic spirit—to, in a phrase, *écrasez l'infâme*.[37] To Voltaire, the best and most practical alternative in France remained a legal, rational, absolute state guided by philosophic principles and supported by a ministry of savants and philosophers.

Voltaire, therefore, continued to act as a leading champion of enlightened absolute monarchy throughout the 1760s and 1770s. He praised Louis XV's *séance de flagellation* before the Parlement of Paris in 1766, which reiterated the principles of absolute monarchy and rejected the sovereignty of the parlements.[38] With the *Histoire du parlement de Paris* (1769), Voltaire defended the heavy-handed attempts of Chancellor Maupeou to tame the parlements. In straightforward prose, he blamed the parlements for the religious and civil wars, derided their hypocritical defense of petty self-interest in the name of the public good, attacked their claims of legal jurisdiction, and castigated their backwardness, intolerance, superstition, and irresponsibility.[39] Later, Voltaire exulted in Maupeou's coup d'état of 1771, which exiled the Parlement de Paris, abolished venality of office, and established six new "superior councils" to administer justice.[40]

Voltaire had little time for exultation, however. Seemingly almost everyone except Voltaire feared and hated Maupeou. Many other philosophes worried about Maupeou's strong-arm tactics and threats to tighten control of censorship.[41] Voltaire's most radical friends and followers, notably Condorcet and d'Alembert, seconded his defense of Maupeou's attack on the parlements, but many others distrusted Maupeou's motives and feared the ultimate results of his reforms.[42] They saw a bully lording over those who claimed to be the noble defenders of the public interest, or at least an overt example of royal despotism. With hundreds of libels and *Maupeouana*, the parlements carried out a successful propaganda campaign painting Maupeou and his associates—not entirely without cause—as corrupt, arbitrary, and despotic. The government, apparently indifferent to public opinion, made little attempt to justify its policies; its best, but hardly welcome, propagandists were Voltaire and his small faction of philosophes. When Louis XV unexpectedly fell ill and died from smallpox on May 10, 1774, the Maupeou Triumvirate—consisting of Maupeou and his associates, the abbé Terray, Comptroller-General of Finances, and the duc d'Aiguillon, Minister of Foreign Affairs—was soon dismissed, and the parlements were recalled.

This was the political world into which Anne-Robert-Jacques Turgot (1727–1781) entered as Comptroller-General of Finances, arguably the single most important office in the kingdom, in 1774.[43] His was a world collapsing from a crisis of legitimacy and foundering under economic strain from massive war debt, ineffective tax policies, and poor industrial development. Having already made a name for himself as an able and innovative administrator while intendant of Limoges from 1761 to 1774, Turgot had earned the confidence of the new chancellor, the comte de Maurepas, who appointed Turgot first as Minister of the Navy, and by late August 1774 as Comptroller-General of Finances. To Voltaire and like-minded philosophes, it seemed that an *esprit supérieur*, as Voltaire called him, had arisen to the highest levels of the government, thus bringing the philosophic dream of enlightened, scientific government close to reality. Voltaire and Condorcet were ecstatic.[44]

Turgot, son of an ancient Norman family that had enjoyed a spotless record of distinguished service to the crown since the sixteenth century, was a *politique* both by blood and by principle.[45] His father, Michel-Étienne Turgot, was renowned as an enlightened official, most noted for serving as *Prévôt des marchands* in Paris and as president of the king's *Grand Conseil*. He served the king faithfully until the end of his life. As Michel-Étienne's youngest son, Anne-Robert-Jacques showed no less promise, and he demonstrated an even sharper intelligence; an inquisitive disposition; a prodigious, practically eidetic, memory; and a complete inability to believe his opinions were ever in the wrong.

As a younger son, he had been chosen for an ecclesiastical career. Dissatisfied with the clerical life and eager to follow in the footsteps of his father and older brother, however, he abandoned the clergy in 1751 and resolved upon a career in the administration. Although there is no reason to assume any particularly anti-clerical or anti-religious sentiments on Turgot's part, there is every reason to believe that Turgot had imbibed the royalist views common to his family. Turgot was also very interested in the new sciences popularized by Voltaire. As a student, Turgot studied the scientific and philosophical works of Newton and Locke, an interest instilled partially by his gifted teacher, the abbé Signorgne at the Collège Duplessis, and partially by his acknowledged favorite author, Voltaire. He corresponded briefly with the great naturalist, Buffon, and many notes survive of his early musings on the arts and sciences. One of his discourses at the Sorbonne, shortly before abandoning his ecclesiastical career, lauded the progress of the arts and sciences in recent times as a force for order, civilization, and morality long suppressed in the Dark Ages that saw, he wrote significantly, "kings without authority, the nobility unchecked, and the people enslaved." In his historical analysis, Turgot attributed the dissipation of the darkness to the gradual recovery since the Middle Ages of towns, the natural commercial, creative, and intellectual centers of any state, and their alliance with monarchs to throw off the burden of "feudal government" that had unnaturally "kept them low." The common interests of kings and towns, Turgot argued, had allowed the rebirth of the arts and sciences that had hastened the "century of Louis, century of great men, century of reason."[46]

Such "philosophic" ideas, while not necessarily inimical to the ecclesiastical life, decidedly pointed Turgot toward royal service. When his friends sought to dissuade him from abandoning the clergy, pointing out that a younger son of a Norman house did not have the fortune to rise to high office in the administration, Turgot replied, "In what you say there is much truth. Take to yourselves the counsel you offer me, since you can follow it. I respect you all, but cannot fully appreciate your view of this matter. As for me, I cannot possibly resign myself to wear, all my life, a mask over my face."[47]

From 1752 to 1760, Turgot held office as *maître des requêtes*. When the Parlement of Paris was temporarily exiled in 1753, Turgot served on the *Chambre Royale* established to hear criminal and civil cases in the absence of the regular court judges. This earned him the lifelong enmity of the parlementary magistrates but was consistent with his expressed political views. During these years, Turgot moved among the intellectual circles of Paris, attending the salons of the famous salonnières, Graffigny, Geoffrin, and Lespinasse, where he met the likes of d'Alembert, Condorcet, Melchior Grimm, and the baron d'Holbach. He was also a frequent visitor at the home of the physiocrat François Quesnay, where he met Adam Smith and, so we are told, mastered the most advanced theories of political and economic science of the time.

D'Alembert in particular took interest in Turgot and drafted him into the ranks of the encyclopedists, for whom he wrote various articles on grammar, metaphysics, physics, economics, and political science. His 1757 article "Fondation" applied economic theories, particularly physiocratic doctrine, to justify reducing the power of privileged corporations in a state. His conclusions strongly criticized arguments in favor of noble autonomy, and in retrospect he had possibly revolutionary interpretations. The reason, he noted, why misery is so common when resources are so abundant is that men living freely off the backs of others not only make the condition of the idle nobleman preferable to the laborer but also "diminish for the State the sum of labor and the production of the earth." At one time conditions may have rationally justified the establishment of such privileged corps (for example, reasons of military defense), but their continued existence could not be justified merely by tradition: "The public utility is the supreme law, and ought not to be compromised either by a superstitious respect for what one calls the *intention of the founders*, as if the ignorant and limited men of one time had the right to enchain unborn generations to their capricious wills, or by the fear of stepping on the supposed rights of certain social bodies." Turgot's recommendation, therefore, was to destroy such foundations when they outlived their usefulness: "If every man who has ever lived had a tomb, it would have been necessary, in order to find land to cultivate, to overturn these sterile monuments and rake up the ashes of the dead in order to nourish the living."[48]

Diderot and d'Alembert were both pleased with Turgot's ability to express radical reforms so reasonably and tried to get more articles from him in 1759, but that was the year in which the Parlement of Paris condemned the *Encyclopédie* (doubtless Turgot's articles did not improve the magistrates' mood). Further volumes did not appear until 1765.[49] By then Turgot was deeply engaged in other activities and wrote no more articles. Later remarks (for example, "I am not an encyclopaedist because I believe in God") denying his involvement with the *Encyclopédie* should not necessarily be taken as a serious rejection of the *Encyclopédie* or its editors, however. Such statements may have been cynical and politically expedient.[50] Even d'Alembert, after all, tried to extricate himself from the project when persecution became too intense.[51] Turgot, hoping to succeed in the royal bureaucracy, did not have the freedom of Voltaire to write what he pleased—and even Voltaire occasionally panicked when confronted with the repercussions of his own audacity. Turgot's actions revealed his clear support for sweeping social and political reform.

In the fall of 1760 Turgot traveled to Geneva and was briefed by his favorite author face to face. D'Alembert's letter of recommendation introduced Turgot as "full of philosophy, enlightenment, and knowledge, and a very good friend of mine."[52] We have no record of Turgot and Voltaire's conversation, but Voltaire was impressed. "I did not know he had written the article *Existence*," Voltaire afterward wrote to d'Alembert. "He is even better than his article."[53] We should not underestimate Voltaire's influence on younger, like-minded reformist thinkers such as Turgot. In the 1760s Voltaire was at the height of his power, and he kept in close contact with those friends he hoped would serve the philosophic cause as he defined it. When Turgot wrote in 1761 to inform Voltaire of his offer of appointment as intendant of Limoges, complaining that he would prefer to find happiness pursuing the consolations of philosophy as a private citizen, Voltaire replied, "I believe that an intendant is the only person who can be useful. Can he not have the highways repaired, the fields cultivated, the marshes drained, and can he not encourage manufacturers?"[54] The lesson was clear: For the Voltairean faction, philosophy was an activity one performed in a public arena. The gardens they cultivated were not private.

It is not necessary here to detail all of Turgot's activities during his thirteen-year stint in the backwater *généralité* of Limoges except when such activities elucidate the application of Turgot's political philosophy, which is perhaps best termed *despotisme légal.*[55] In short, Turgot set out with great industry to follow Voltaire's suggestions. This involved rationalizing the administration of the *généralité*, forming a civil service, completing a land register and taxability study, reforming the onerous highway labor requirements (the *corvées*) as well as the militia, and opening the economy to free trade.[56] For the most part, Turgot used negotiation and education to gain support for his reforms. The sciences were his allies, not merely as tools but as ideological support: he employed the Limoges Society of Agriculture to propagate physiocratic doctrines, to encourage research on the storage of grain, to improve plowing practices, to eliminate cattle diseases, and to introduce new staple crops such as the potato. He also promoted the publication of works on free trade and distributed copies of Guillaume François Le Trosne's physiocratic work *La Liberté du commerce des grains toujours utile et jamais nuisible* (1765), which gave the arguments for free trade in short form. His reforms were, eventually, entirely successful and won him the respect and gratitude of the people of Limoges. When he departed Limoges in 1774 to become Minister of the Navy, the peasants left their fields to bid him farewell and the priests said masses in his honor.[57] He left the province with a better economy than he had found there and with the best roads in the kingdom.

Turgot may have believed that his rational reforms, justified by the natural and economic sciences, would naturally sway a public skeptical of change if only the benefits were explained clearly, but in truth many of his accomplishments required a heavy hand and ran roughshod over the opposition. His reform of the *corvées*, for example, reimbursed the loss of those parishes most affected—those near the roads—with a tax break that took the form of real payments—that is, laborers were actually paid to work. This meant raising the taxes in less affected parishes by an equivalent amount, thus spreading the burden of road construction more fairly over the whole *généralité*. When aristocrats and bourgeois organized opposition that resulted in some local courts voting against the reform, Turgot simply ignored their decisions. If men showed up to do roadwork, they were paid, and the reform was a fait accompli. Showing his scorn for the magistrates in a letter to his friend Daniel Trudaine, the director of roads and bridges, Turgot wrote, "Up to the present the Parlements have always sought to show themselves as defenders of the people against the Royal Court, and the Court would do well to overturn this view and prove that the Parlements are opposed to the well-being of the people."[58]

Perhaps the most striking example of Turgot's willingness to administer his province in an enlightened but nevertheless dictatorial fashion was his vigorous enforcement of the new free trade laws of 1764, which he had helped to draft. Under the reform-minded Comptroller-General of Finances, François de l'Averdy, Turgot and Trudaine put together a new legislation that would finally end the ancient protections and restrictions that inhibited the free flow of grain throughout the kingdom. Physiocratic theory held that restrictions in trade actually exacerbated the local famines they were designed to prevent. While one locality experienced dearth, excess grain rotted in another. The attempts by the central government to organize food relief, however, tended toward corruption, price speculation, and general ill will. Free trade, physiocrats argued, should spread the grain supply more evenly over the entire kingdom without direct government involvement. Unfortunately, a poor harvest in 1765 (an all-too-common occurrence) caused a worrisome price increase, leading to grain riots and general alarm. The parlements played on the fear of the populace in order to organize resistance to government policy. Louis XV, determined not to lose authority to the magistrates, upheld the reasoning of the free traders, who argued that it was not free trade itself but the failure of the harvest that led to the price increase.

To head off opposition in his own province, Turgot first resorted to education and persuasion, seeking, according to his preferred method, to justify the law by an appeal to reason and the sciences. He posted the laws throughout Limoges, distributed more copies of Le Trosne's *La Liberté du commerce des grains toujours utile et jamais nuisible*, and ordered his civil service to listen patiently to the people's complaints against the new law and to respond with the "language of reason not of authority."[59] "It is absolutely necessary," explained Turgot, "that the grain trade is free. If the inhabitants of one town take it upon themselves to inhibit the transport of grain, others will believe they have the same right, and those regions where the dearth is greatest, without aid, will die of hunger."[60]

Where persuasion and education failed, however, Turgot resorted to force. He ordered his men to take "all means that an enlightened conscience suggests" to stop those who would foment unrest or resistance. This included the dispersion of "seditious crowds," the arrest and prosecution of anyone interfering with the free flow of grain, and the use of spies to watch over the towns and to call for police reinforcements if necessary. Turgot had little time for those who opposed his authority, and he grew increasingly angry about the obstructionism of the parlements. When the Parlement of Paris forbade free trade in 1768 and the ministry, in sudden weakness, failed to silence or oppose the Parlement, Turgot inveighed furiously to his friend, the physiocrat Pierre Samuel du Pont de Nemours, "I detest the barbarity of the *boeuf-tigres* [a term of contempt for the magistrates] and I scorn the poltroonery of those [the ministers] who allow themselves to be kicked and gored by these stupid but ferocious animals."[61] Turgot himself refused to lie down. When the weakness of the ministry led to the dismissal of Maynon d'Invault (l'Averdy's successor and also a supporter of free trade) and the appointment of the abbé Terray, Turgot resolved to uphold free trade in his jurisdiction at all costs. He ignored orders from Terray to reinstitute price controls and prosecuted anyone who hindered merchants in his territory. Eventually Terray allowed his headstrong subordinate to have his own way, and Turgot successfully organized famine relief that did not oppose the principles of free trade. Defiantly he wrote to Terray, "There are circumstances when the wisdom of the chief administrators is forced to allow a complete liberty on the manner of execution to those who are charged with it and who are actually on the ground."[62]

In the backwater province of Limoges, Turgot had power and could impose his will as a de facto enlightened despot with little fear of reprisal. Matters were different when he was recalled to Paris in 1774 to serve, briefly, as Minister of the Navy and, also too briefly, as Comptroller-General of Finances. The weaknesses of the monarchy concentrated at the top. During his short stint as Minister of the Navy, from July 19 to September 8, 1774, Turgot was generally ignored, it is true, and his successors simply abandoned his most radical reforms. His thirty-five days at the head of the navy, however, were a windfall for the scientific community, especially to the Royal Academy of Sciences. Mostly at the instigation of his young friend, Condorcet, Turgot ordered various projects of scientific utility, including a new translation of Euler's *Théorie complète de la construction et de la manoeuvre des vaisseaux*, the introduction of Magellan's naval instruments into regular use, desalination experiments directed by the famous chemist Antoine Lavoisier, and royal permission for the botanist Saint-Edmund to mount an expedition to Peru.[63] There was little opposition to such projects because everyone agreed they were useful and because they did not attack any economic or political special interests. The projects had a subtle ideological goal, however: active state support for practical scientific innovation and experimentation to raise the government's image as rational, not arbitrary, in the public eye. As we shall see, Turgot was quite aware of the need to improve the image of the monarchy in 1774. On a more personal level, Turgot's employment of the scientific community enhanced the funding and prestige of the Royal Academy of Sciences, thus earning Condorcet, who had used his influence with Turgot to get the projects funded, the gratitude of the academicians who unanimously elected him *secétaire perpétuel* in 1776. One cannot ignore favoritism as an aspect of the role the sciences played in Turgot's ministry, but such preference does not negate the ideological objective behind it: to legitimize state policy by associating it with scientific authority. In the achievement of this goal, the happiness of the scientific community mattered a great deal.

Turgot's more radical proposals for the navy would have controversially imposed the principles of free trade on the naval industry. In what amounted to a complete reversal of Colbert's ancient but still active policy of mercantilism, Turgot proposed to transfer most shipbuilding contracts to Sweden, where plentiful wood and iron made it much cheaper to build ships. Labor and supplies would be Swedish, but French engineers would supervise construction. The state would save in construction costs while also providing cash to the Swedes to buy French wines, coffees, clothing, and other luxury products. Turgot's attempts to open the French economy, however, attacked privilege in virtually all sectors, and his proposals for the navy would have put French shipwrights, laborers, and suppliers out of work. His successors, of course, let the matter drop.

Initiating reform in the navy was child's play compared to reforming the kingdom's finances, however, and Turgot harbored few illusions about his chances of success or the sacrifices success would require. In his letter to Louis XVI accepting the post of Comptroller-General of Finances on August 24, 1774, he reminded the young king tersely of the steps necessary to improve the credit of the government: there must be no bankruptcy, no increases in taxation, and no borrowing. "To fulfill these three aims there is only one means," he wrote. "That is to reduce expenditure below that of receipts, and enough below to save each year twenty million [*livres*], in order to pay off old debts. Failing this, the first gunshot will force the State into bankruptcy." Turgot sensed furthermore what the reaction of the privileged classes would be to his frugality, and he begged the king's absolute trust:

> I have felt the value of the confidence with which you honor me; I have felt that you entrust me with the happiness of your people, and, if I may say so, the job of promoting the love of your person and authority. But at the same time I

> have felt the danger to which I expose myself. I foresee that I will be alone in fighting against abuses of every kind; against the efforts of those who profit from these abuses; against the crowd of prejudiced people who oppose all reform, and who are such powerful instruments in the hands of those interested in perpetuating disorder. . . . I will be feared, hated even, by most of the Court, and by all who solicit favors. They will attribute all refusals to me and will paint me as a hard man because I advise Your Majesty not to enrich even those that you love at the expense of your people's sustenance.[64]

Turgot was no idealistic fool who believed he could force through major reforms with impunity as he could in Limoges—although he may have underestimated the tenacity of his opponents, who fought him to the bitter end. From a realistic perspective, however, there was likely no alternative to his method other than a revolution. In brief, his method for reforming the government by changing its system of finances involved only a slight modification of his usual method: negotiation and education followed by force. Turgot's plan was to improve the credit of the government, both financially and morally, to such a degree that the privileged classes, in the face of public support for a rational, efficient, and effective administration, would have no choice but to acquiesce to reform.[65] Turgot's method relied heavily on the cooperation of the scientific community. To the delight of those who promoted the idea of the sciences as morally enlightening, the sciences were finally drafted into the business of active government reform. Improving the financial and moral credit of the monarchy required a great deal of popular confidence in a government long vilified as excessively bureaucratic, arbitrary, and inefficient, so Turgot called on savants, whose public reputation had risen considerably in the eighteenth century, to provide rational, scientific authority to his reforms. His efforts to cut expenditures and open the economy to free trade relied everywhere on scientific expertise.

We find a perfect example in the attempt to unify France's system of weights and measures.[66] Unrestricted internal trade depended upon a universal standard, but the provinces and regions used a dizzying morass of incompatible systems. A universal standard, free trade advocates argued, would tend to promote internal trade without continual government interference. Turgot, therefore, named Condorcet and another academician, Mathieu Tillet, as *inspecteurs des monnaies* (inspectors of the mint) in early 1775 to solve the problem. Rather than merely adopt as a standard one of the many irrational systems that had grown haphazardly from practice and convention, Condorcet proposed to do away completely with all the old units of measurement and establish a new, scientifically exact system based on a geophysical constant: the length of the pendulum swing striking the second at sea level, 45 degrees latitude. Turgot accepted this proposal, and in late summer Condorcet informed Turgot that the naval astronomer, Charles Messier, would be on site near Bordeaux to take the measurements.[67] Turgot's successors abandoned this plan, but Condorcet finally achieved the unification of weights and measures during the revolution in 1792.[68]

Basing a system of weights and measures on a scientifically determinable standard had the dual benefit of associating the government with rational instead of arbitrary reform—as the choice of, for example, the Parisian system as a standard would have—and of providing the scientific community with practical work and funding, thus more closely associating it with the government.[69] Had the plan worked, the new system would also have contributed to opening the economy of Old Regime France.

The project to standardize weights and measures was by no means Turgot's only plan to associate scientific authority with free trade reform. In an effort to open France's interior waterways to shipping, and thus to break down further barriers to free trade, he ordered in late summer and fall of 1775 canal constructions in Picardy and Bourgogne, and public works to make the Charente River navigable between Angoulême and Cognac. In September, he commissioned three new offices, the Inspectors-General of Navigation, filled by the academicians d'Alembert, Condorcet, and Charles Bossut, and he earmarked 800,000 *livres* to create a general plan for interior travel and navigation. The inspectors journeyed to Flanders and Picardy in October to study canal construction and hydraulic science. Eventually a new chair of hydraulics, to be held by Bossut, was established.[70]

In November, Turgot ordered the Academy of Sciences to establish a commission to study means of curing an epidemic hurting the cattle industry in Navarre and Guienne. The commissioners failed to discover a cure, having no understanding of microbes or antibiotics, but they obtained permission to quarantine and exterminate infected animals, which curtailed the spread of outbreaks. In 1778, the commission, still at work, was given formal status as the *Société royale de médecine*.[71]

More ominous to vested interests was Turgot's reform of the gunpowder industry. As part of his effort to improve government revenues, Turgot made efforts to replace the private government-granted monopolies, or *fermes*, with government administrations, or *régies*. In May 1775, upon the advice of the Academy of Sciences, Turgot revoked the contract of the notoriously corrupt gunpowder monopoly, the *salpêtriers du Roi* and their underlings, and replaced it with a new *régie* under Lavoisier's administration, with orders to engage in research on more efficient saltpeter production. The new methods were eventually so successful that Lavoisier wrote in 1789 that they were saving the state 2,400,000 *livres* annually, as well as producing the best gunpowder in Europe.[72]

Such enormously successful and scientifically justified reforms, combined with Turgot's efforts to economize by suppressing, reforming, or regularizing corrupt forms of direct and indirect taxation, began to improve drastically the credit and image of the government as an efficient, rational, and financially solvent institution. By the end of 1775, Turgot was actually balancing the state budget, which improved credit enough that he was capable of negotiating new and better-financed loans to pay off older, less favorable ones.

Turgot also hoped to improve the image of the government as moderate and open to compromise, even if in truth it was not. Maupeou's suppression of the parlements in 1771 had been unpopular, despite Voltaire's wholehearted support. One of Turgot's first acts as Comptroller-General, therefore, was to advise the king to allow the Parlement of Paris to return from exile.[73] Unfortunately, efforts to work with the parlements, which had little interest in compromising with Turgot, were bound to fail. The parlements had much to fear in a popular, functioning centralized government, since, as its financial and moral credit improved, Turgot might be tempted to carry out much more radical reforms that would endanger noble privilege. When Turgot desired new loans to pay off bad debts, therefore, he could arrange no deals with French financiers (allied with the parlements, they would rather artificially engineer a shortage than give a loan to Turgot), and he was forced to negotiate with Dutch bankers.

Turgot inspired a much more violent reaction when he reenacted the free trade laws of 1764. By sheer bad luck, a mediocre harvest in 1774 caused the price of grain to rise once again. Financiers and monopolists, who benefited least from free trade, incited popular protest and engineered the infamous "Flour War," probably supported financially by one of the strongest opponents of absolute monarchy, the Prince de Conti.[74] In March and April, riots spread from Dijon to Versailles and Paris. To critical observers, it quickly became apparent that the riots were not spontaneous outbursts by mobs of starving peasants demanding bread. The rioters were well fed and armed and supplied with intelligence on the location of grain stores, indicating that considerable planning and financing went into their operations. The "rioters" moved with precision to destroy grain stores, as no starving peasant would have done, with the obvious goal of forcing prices higher in order to discredit Turgot's free trade laws.[75] On May 12, the abbé Morellet, who had observed the disturbances, remarked with excessive caution, "Without believing that there was to all this a first and unique cause, a plot formed and directed to a single end, one cannot disguise the fact that the first movements were carefully held together. . . . We have been above all strangely surprised about the ease with which Paris was pillaged."[76] Those closer to Turgot did not hesitate to accuse his enemies of plots. Condorcet particularly suspected Jacques Necker, the ambitious and wealthy Swiss banker who had published *La legislation et le commerce des grains* in March 1775, which slandered Turgot's economic policies. Necker probably did not have the capability or resources to organize the Flour War, but he may well have been associated with those who did. Nevertheless, in May 1775 Condorcet published a violent polemic against Necker, *Lettre d'un laboureur de Picardie à M. Necker, auteur prohibitif à Paris*, followed in June by the officially sanctioned *Réflexions sur le commerce des blés.*[77] In the *Letter*, Condorcet accused the rich banker of encouraging the riots. He also scorned the bad faith and self-interested behavior of the rich, who lived at the expense of the poor. The letter was so accusatory that it earned Condorcet the ire of his more moderate friends, some of whom had financial dealings with Necker.[78] Voltaire, of course, praised Condorcet for talking straight: "This ought to redirect their minds toward what little there still is in Paris of good sense and good taste."[79]

If the goal of the Flour War was to discredit Turgot's reforms and remove him from power, it only partially succeeded. Libel poured forth against him, but Turgot remained firm and refused to contemplate revoking the free trade laws. Perhaps most importantly, he retained the trust and confidence of the king. Louis XVI upheld Turgot's laws in the face of his courtiers, who demanded they be repealed. At the height of the disorder, he gave Turgot carte blanche to deploy the military. Turgot did not hesitate. He moved ten thousand troops to Versailles to attend the king and twenty-five thousand more into Paris. In short order he restored peace and established an extraordinary *commission prévotale* to try arrested insurgents. The Parlement of Paris, fearful lest the *canaille* get out of hand, supported the imposition of order. Still eager to discredit the free trade laws, however, it ruled that the king ought to do something to reduce the price of bread. When the magistrates tried to post this ruling publicly, Turgot had his troops smash their printing blocks and pull down their placards. He then ordered the king to hold a *lit de justice* to register his new criminal tribunal. The king wrote from Versailles on May 6, "I have just carried out, Monsieur, what we agreed on." Reveling, perhaps, in his newfound power, the king wrote again the next day, "They will see from this that I am not as feeble as they believed and that I know how to execute my resolutions, which will make it easier for you to do what must be done."[80]

In the short term the results of the Flour War appeared positive for Turgot and the proponents of enlightened absolute monarchy. In one move, Turgot had become the right-hand man of the king, rising even above Chancellor Maurepas in influence. The Parlement of Paris, although still defiant, had bowed to Turgot's will in the king's name, and martial law had restored order in the capital. The manner of Turgot's success, however, revealed the fundamental flaw of his methods. The vested interests of the kingdom had not backed down in the face of a more popular and credit-worthy government, and Turgot's attempts to justify his reforms by an appeal to scientific reason had failed. Force had been his only recourse.

Turgot's mastery of the king's will had not boded well, either. Despite his own protests to the contrary, Louis XVI was indeed feeble, easily influenced, and lacking in self-confidence.[81] Turgot's power was tied to a fundamentally weak-willed and unreliable master. His enemies well knew that the key to their success was to drive a wedge between Turgot and the king.

Turgot's final and most radical acts as Comptroller-General only demonstrated further the weaknesses of both the government and Turgot's methods of carrying out reform. Inspired by his dramatic success in the Flour War and by his new credit with the king, Turgot issued in early 1776 his famous Six Edicts. Although the edicts, and the ideas behind them, were part of a long-term project, Turgot clearly felt he had to move forward while he still held the unconditional confidence of the king. In short, the edicts ordered the opening of the economy to free trade and attacked privilege and corruption. They suppressed all duties levied on grains and cereals and abolished the *Caisse de Poissy*, thus freeing the meat and cattle market from price controls. The edicts also suppressed the multiplicity of corrupt officials at the ports, quays, halls, and markets of Paris and abolished the guilds and the *corvées*. In eliminating the *corvées*, the edicts also thoroughly condemned privilege and set the stage for heavier taxation of privileged landowners. In his report to the king, Turgot sought to prepare Louis XVI's mind for the resolve that would be required to succeed in this revolutionary venture by reminding him that the *corvées* were a great injustice that benefited the privileged at the expense of the poor. Turgot understood that, in the end, attempts to impose further taxes on the aristocracy under the current regime would be accomplished by force, not reason:

> Once the suppression of the *corvées* is decided upon, it is impossible not to replace them with an imposition in taxes, because it is necessary that the roads be built.
>
> Your Majesty understands the justice of imposing this tax on the landowners: it is therefore on them that it must be established.
>
> This imposition will be difficult, but when a thing is recognized as just, when it is an absolute necessity, one cannot stop because of some difficulties: one must vanquish them.
>
> The first of these difficulties is the repugnance that in general the privileged have of submitting to a tax, new for them, that the taxable have up to now supported alone.
>
> All those who deliberate on the registration of a law [i.e., the magistrates] are privileged, and one cannot fool oneself that they are all above personal interest, even if this particular tax does not extend to all of them. It is reasonable that this motive will secretly influence a great part of the objections that will be made. I would not be surprised even if some publicly avow this motive, nor if some find specious scholarly reasons to color their objections. The solution to this difficulty rests in the justice of Your Majesty and in his firm will to have done what his justice tells him.[82]

Louis XVI immediately approved Turgot's edicts, and they received enthusiastic support from the philosophes, particularly Condorcet and Voltaire.[83] With predictable doggedness, of course, the magistrates began their offensive even before receiving the texts from the king's Council of State. The magistrates' ally and informant on the council, the marquis de Miromesnil, Keeper of the Seals, argued their case in debates with Turgot—debates that revealed the impossibility of compromise. Asserting the popular thesis that the nobility must be exempt from taxation because it was entrusted with the military defense of the kingdom, Miromesnil warned of the dangers of abolishing privilege: "I cannot say too strongly that in France the privilege of the nobility must be respected, and that it is, I believe, in the interest of the king to maintain it." Turgot found such arguments absurd and responded with exasperation that unless a tax was merely tribute imposed forcibly by the strong upon the weak, then, in a just society, all must pay their fair share: "The expenses of the government having as their object the interest of all, all must contribute to them; and the more one enjoys the benefits of society, the more one ought to hold it as an honor to partake of these charges. From this point of view it is difficult to see the pecuniary privilege of the nobility as just."[84]

Monarchy and parlements were therefore at an insurmountable stalemate when the edicts were finally submitted on February 7, 1776. The Parlement of Paris, led by the intractable Prince de Conti, remonstrated on March 4, and the king, in anger, demanded they register the edicts on March 6. The Parlement remonstrated again the next day. Meanwhile, agents of the opposition continued a flow of libel against Turgot, conspired to cause shortages of bread, sought to influence the king, and incited strikes.[85] In the Council of State, only Turgot and his friend and ally, Lamoignon de Malesherbes, Minister of the Household, advised Louis XVI to continue opposition to the parlements. Scheming, Maurepas had chosen to withdraw his support for Turgot and say nothing either way. Still placing his confidence in Turgot, the king held a *lit de justice* on March 12 and forced the Parlement to register the edicts. Once again Turgot succeeded only by resorting to force—but force would never quell the opposition. Enlightened absolute monarchy, and what the Voltairean faction considered the rule of reason, was proving a failure.

It is unnecessary here to recount in detail the means by which Turgot's opponents finally broke Louis XVI's confidence in him and brought about his dismissal in May: the tale abounds with libel, assaults, petty crimes, lies, and forged correspondences containing treasonous statements purporting to be between Turgot and his friends. Louis XVI may also have begun to fear and resent the manner in which Turgot seemed to encroach upon royal authority and was thus willing to believe the lies. "M. Turgot wants to be me and I do not want him to be me," the king is said to have once famously remarked.[86] As Louis XVI withdrew his support, Turgot wrote to him in disbelief, "I cannot conceive, Sire, that from levity of heart you will consent to sacrifice all your reign and the welfare of your people: it would be necessary to think that Your Majesty has not believed a single word I had said or written."[87] Despite Turgot's appeals, the king demanded his resignation on May 12. Malesherbes, in frustration, resigned the same day. Those who succeeded them, including Turgot's enemy, Necker, as director of the treasury, immediately set about reversing Turgot's reforms. As the guilds and *corvées* were restored, Turgot began, in despair, to write a letter to Chancellor Maurepas, incredulous that the chancellor could support such a thing: "I want therefore to try and remind you of yourself, of what you have said and thought a thousand times, of what you owe to the public, to the King, to your own reputation, to oppose your conscience to passion which, permit me to tell you, hides from you your true interests."[88] The letter was never finished, nor sent, and Turgot in quiet despair retired from public life to, at long last, cultivate his own private garden.

For the Voltairean faction, the fall of Turgot marked the beginning of the end of the absolutist monarchy as a political ideal. The monarchy could not achieve the reforms Voltaire and his associates desired because its only recourse was to a degree of force it neither could nor would sustain. The appeal to scientific reason under the monarchy was, arguably, a failure—a failure that brought the members of Voltaire's faction to fury and despair, not least because the moral idea of the sciences had, for them, become hardened almost into an inflexible ideology. Anyone who opposed their ideas for reform was the enemy. For Voltaire—whose despair was much less quiet than Turgot's—there were no realistic options left. In September he wrote morosely to Condorcet: "Since that fatal day, I have not followed anything, I have not requested anything of anybody, and I am waiting patiently for someone to cut our throats."[89]

To Condorcet, the fall of Turgot in May 1776 was a disaster, a catastrophe for the goal of enlightened reform under the monarchy, and he confessed that the subsequent celebrations at Versailles by those people whom he could only define as ignorant and misled broke his heart.[90] Perhaps because of his youth and his closeness to Turgot, Condorcet reacted even more melodramatically than the other philosophes. He saw the event not merely as the failure of a desired plan for reform, nor only as the disgrace of a friend, but as the loss of a personal opportunity to engage in public service for a perceived greater good, for a *cause philosophique*. In early May he complained to his friends, the Suards, "It is sad to no longer work but for one's own pleasure . . . when one had, for some time, the hope of working for the good of men."[91] He expressed the same sentiments more dramatically later in a letter to Voltaire: "I have not written to you . . . since the fatal event that banished hope and courage from all honest people. I have waited until my anger was passed and there only remained the pain. This event has changed all of nature for me. I no longer have the same pleasure while gazing on this beautiful country where happiness was born. . . . See how far we have fallen, my dear and illustrious master, and from so high! Adieu. . . . We have had a pretty dream, but it has been too short. I will return to geometry and philosophy. It is a cold thing no longer to work but for vainglory when one flattered oneself for some time that one worked for the public good."[92]

Although perhaps sentimental and melodramatic by contemporary, more cynical standards, Condorcet's anger and despair apparently had an effect on his understanding of reform and on the relationship between the sciences and the state. The moral idea of the sciences as Condorcet had come to formulate it called for the establishment of order through the agency of savants under a strong state. The failure of Turgot, therefore, dashed the hope that the monarchy could form that strong state, but Turgot's failure did not at all convince Condorcet that the idea of the sciences was in any way wrong or impossible. As the next chapter will explore, the fall of Turgot precipitated the radicalization of Condorcet's political views. After 1776, Condorcet abandoned the idea that reform could proceed from the monarchy; he began to become a revolutionary.

NOTES

1. Jonathan Israel traces the rift in the *parti philosophique* between "Voltairean" (or Newtonian) and "radical" latter-day "Cartesians." See Israel, *Enlightenment Contested*, "Realigning the Parti Philosophique: Voltaire, Voltairianisme, Antivoltairianisme (1732–1745)," 751–80.

2. See Sutton, *Science for a Polite Society*.

3. See James E. McClellan III, *Science Reorganized: Scientific Societies in the Eighteenth Century*, especially chapter 3, "The Scientific Society Movement to 1750," 67–108; and Daniel Roche, *Le siècle des lumières en province: Académies et académiciens provinciaux, 1680–1789*.

4. On arguments for (*thèse royale*) and against (*thèse nobiliaire*) strong monarchy, see Wright, *A Classical Republican*, and also Peter Gay, *Voltaire's Politics: The Poet as Realist*, chapters 2 and 7.

5. Critiquing the application of the Weberian concept of bureaucracy as rational, Keith Baker shows that the eighteenth-century use of the word was often pejorative, and it indicated arbitrary and unaccountable government. See Baker, "Science and Politics at the End of the Old Regime," in *Inventing the French Revolution*, 160–63.

6. See, for example, the marquis d'Argenson, *Considérations sur le gouvernement ancien et présent de la France* (1739). D'Argenson, much like Sorbière (see chapter 1), argued that a monarchical government is not despotic, and in fact is more conducive to personal liberty than an unstable aristocratic constitution because there can be no liberty without political order. Quesnay and the physiocrats used the term *legal despotism*. See Shovlin, *The Political Economy of Virtue*, 107–8.

7. Peter Gay, *The Enlightenment: An Interpretation*, vol. 2.

8. Of course, opponents of absolutism argued that it was the constitutional privileges and liberties of the traditional orders that prevented the monarchy from slipping into despotic excesses and so checked *its* use of bureaucratic, arbitrary, and unaccountable power. They found support in the works of philosophes (e.g., Montesquieu) who stressed tradition and history—or pseudo-history, often myths about medieval France—rather than science and mod-

ernity, as the basis for moral reform. See Smith, *Nobility Reimagined*, and Wright, *A Classical Republican*.

9. Charles Gillispie, *Science and Polity*, 550.

10. See, for example, Goodman, *Republic of Letters*.

11. Voltaire to d'Alembert, October 16, 1765, *Oeuvres Complètes de Voltaire, nouvelle édition*, 44: 88.

12. Baker, "Science and Politics at the End of the Old Regime," 166.

13. On the Brittany Affair, see John Rothney, *The Brittany Affair*. See also Maza, *Private Lives and Public Affairs*.

14. Voltaire, *Le Siècle de Louis XIV*, vol. 2, chapter 31.

15. D'Alembert, *Discours Préliminaire*, in *Encyclopédie* . . . (articles choisis), 153. "La philosophie, qui forme le goût dominant de notre siècle, semble par les progrès qu'elle a fait parmi nous, vouloir réparer le temps qu'elle a perdu."

16. For example, see Fontenelle, "Préface de l'histoire de l'Académie des Sciences depuis 1666 jusqu'en 1699," *Oeuvres Complètes de Fontenelle*, 1: 2. "Enfin toutes les sciences et tous les arts dont le progrès était presque entièrement arrête depuis plusieurs siècles, ont repris dans celui-ci de nouvelles forces, et ont commencé pour ainsi dire, une nouvelle carrière."

17. Joining the Academy in 1718 as an associate geometer, Mairan (1678–1771) became a pensioned geometer in 1719. In addition to scientific, mathematical, and literary work, he served as *sous-directeur* of the Academy five times, and director four times. He received his *vétérance* in 1743. On Mairan, see Grandjean de Fouchy, "Eloge de Mairan," in *Histoire de l'Academie Royale des Sciences*, 89–104; Abby Kleinbaum, *Jean Jacques Dortous de Mairan (1678–1771): A Study of an Enlightenment Scientist*; and "Mairan, Jean Jacques d'Ortous de," in Charles Coulston Gillispie, *Dictionary of Scientific Biography*, 33–34.

18. Condorcet, *Oeuvres de Condorcet*, 2: 318–19. "M. de Fontenelle avait . . . qui convenait le mieux au moment où la marche naturelle des esprits avait marqué pour les peuples de l'Europe le passage de la servitude des préjugés au règne de la raison . . . l'un avait à rappeler l'utilité générale des sciences, la beauté de spectacle qu'elles présentent à l'esprit humain, le noble exercice qu'elles offrent à son activité; l'autre [Fouchy], parlant à des hommes déjà passionnés pour elles, devait se borner à faire sentir l'utilité de chaque travail, pour le progrès de la science à laquelle il appartient. L'histoire de M. de Fontenelle devait être plus philosophique; celle de M. de Fouchy plus savante."

19. Fouchy, "Eloge de Fontenelle," 282–306. "[Q]uelquefois un grand homme donne le ton à son siècle. Il a été lui-même ce grand homme qu'il annonçoit, et on peut le regarder comme un de ceux auxquels les Sciences, et par conséquent des hommes, ont le plus d'obligation, et comme modele que ceux qui lui sucéderont, devront toujours s'efforcer de suivre."

20. Attributing the progress of the "philosophic spirit" of the age to Fontenelle had become axiomatic. In his own eulogy of Fontenelle on February 1, 1757, Grimm said, "L'esprit philosophique, aujourd'hui si généralement répandu, doit ses premiers progrès à Fontenelle." Friedrich Melchior Grimm, *Correspondance Litteraire, philosophique, et critique*, 16 vols., ed. Maurice Tourneux.

21. Grandjean de Fouchy, *Eloges des Académiciens de l'Académie Royale des Sciences, morts depuis l'an 1744*, tome premier (Paris, 1761), preface, iii–viii. "Je suis bien éloigné de penser que ceux que je présente aujourd'hui ayent les mêmes titres pour mériter l'attention du Public; mais il sont la suite des premiers, ils tendent au même but, & si on y trouve moins d'art, moins d'éloquence & moins agrémens, au moins puis-je répondre du soin que j'ai apporté à peindre le plus fidelement qu'il m'a été possible, le caractere & les vertus de ceux dont j'ai eu à parler."

22. D'Alembert, "Réflections sur les éloges Académiques," *Oeuvres Complètes de d'Alembert*, 2: 150–53. "Si les anciens qui élevaient des statues aux grands hommes, avaient eu le même soin que nous d'écrire la vie des gens de lettres, nous aurions, il est vrai, quelques mémoires inutiles, mais nous serions plus instruits sur les progrès des sciences et des arts, et sur les découverts de tous les âges, histoire plus intéressante pour nous que celle d'une foule de souverains qui n'ont fait que du mal aux hommes."

23. D'Alembert, "Eloges Académiques," *Oeuvres de d'Alembert*, 4: 535–38, and 2: 150–53. "[S]ans l'art avec lequel ce célèbre écrivain a fait valoir la plupart des ouvrages de ses

confrères, ces ouvrages, quoique excellens, ne seraient connus que des savans seuls; ils resteraient ignorés de ce qu'on appelle *le public*; et la consideration dont jouit l'Académie des sciences serait moins générale. . . . Son genre d'écrire lui appartient absolument, et ne peut passer, sans y perdre, par une autre plume; c'est un liqueur qui ne doit changer de vase."

24. Condorcet, *Histoires et Mémoires de l'Académie royale des sciences*, 1783, "Histoires" 1–4. "On n'a plus besoin de dire aux princes, qu'ils ont intérêt de protéger les sciences; ni au public, que les savants ont des droits à sa reconnaissance "

25. Edwin Van Meerkerk and Theodore E. D. Braun, "From the *Voyage du baron de Gangan* to *Micromégas*: New Documentary Evidence Linking Voltaire's Space Odysseys," *SVEC* 6 (2002): 385–98.

26. Voltaire, *Micromégas* (1752), chapter 2.

27. See, for example, Voltaire to Fontenelle, June 1721 (after reading the *Entretiens sur la pluralité des mondes*), *Oeuvres Complète de Voltaire*, 33: 56–57. "C'est à vous que nous nous addressons, monsieur, comme à notre maître. Vous savez rendre aimables les choses que beaucoup d'autres philosophes rendent à peine intelligibles; à la nature devait à la France et à l'Europe un homme comme vous pour corriger les savants, et pour donner aux ignorants le goût des sciences."

28. See *Voltaire's Notebooks*, ed. Theodore Besterman, 1: 60, 67.

29. On Fontenelle's influence, see Ira Wade, *The Intellectual Development of Voltaire*, 234–36, 428. As J. B. Shank points out, Voltaire carefully managed and manipulated the public perception of science, and especially the reception of Newton's work, in order to create an amenable, and even implicitly radical, scientific culture—a Newtonian Enlightenment—in France. See Shank, *The Newton Wars*.

30. *Histoire de Charles XII* in *Oeuvres Complètes de Voltaire*, 16: 145–354.

31. Voltaire, *Letters on England*, translated by Leonard Tancock, letter 9, 47–50.

32. On the influence of d'Argenson and Dubos, see Gay, *Voltaire's Politics*, 102–9.

33. Voltaire, *Le Siècle de Louis VIX*, 2: chapter 31. "On ne croirait pas que les souverains eussent obligation aux philosophes. Cependant il est vrai que cet esprit philosophique qui a gagné presque toutes les conditions, excepté le bas peuple, a beaucoup contribué à faire valoir les droits des souverains. Des querelles qui auraient produit autrefois des excommunications, des interdits, des schisms, n'en ont point causé. Si on a dit que les peuples seraient heureux quand ils auraient des philosophes pour rois, il est très vrai de dire que les rois en sont plus heureux quand il y a beaucoup de leurs sujets philosophes."

34. The clergy reprimanded Louis XV about attempting to "meddle with God's Property," but in a more practical vein it also warned him, not so subtly, that any attempt to impose further tax on the clergy would result in his inability in the future to secure loans from the Church. Louis got the message and backed down. See Michael Kwass, *Privilege and the Politics of Taxation in Eighteenth-Century France*, 24–33.

35. Voltaire, *La voix du sage et du peuple* in *Oeuvres Complètes de Voltaire*, 23: 466–71 (cited in Gay, *Voltaire's Politics*, 136–37).

36. See George Havens, *Voltaire's Marginalia on the Pages of Rousseau: A Comparative Study of Ideas*, 68. Voltaire's notes concerning Rousseau's "civil religion": "[T]oute dogme est ridicule funeste. Toute contrainte sur le dogme est abominable. Ordonner de croire est absurde. Bornez vous à ordonner de bien vivre."

37. See, for example, the article "Superstition" in the *Dictionnaire philosophique*.

38. See letter to Damilaville, March 12, 1766, *Oeuvres Complètes de Voltaire*, 44: 243. "Vous m'avez fait un très-beau présent en m'envoyant la réponse du roi au parlement. Il y a longtemps que je n'ai rien lu de si sage, de si noble, et de si bien écrit. Si le roi n'était pas protecteur de l'Académie, il faudrait l'en mettre pour cet ouvrage."

39. *Histoire du parlement de Paris* in *Oeuvres Complètes de Voltaire*, 16: 1–109.

40. See Voltaire's note added to the *Essai sur les moeurs* in 1775, *Oeuvres Complètes de Voltaire*, 12: 205. Also see his poem to Maupeou in 1771, ibid., 10: 374: "Mais que Maupeou tout seul du dédale des lois / Ait su retirer la couronne, / Qu'il l'ait seul rapportée au palais de nos rois; / Voilà ce que je sais, voilà ce qui m'étonne."

41. Maupeou's desire to "muzzle the philosophes" was taken seriously. See Gay, *Voltaire's Politics*, 316–17.

42. Voltaire may have distrusted him as well, but he was willing to sign a pact with the devil to obtain his desired goals; as Gay writes, "Voltaire was never particular about his tactical affiliations." Gay, *Voltaire's Politics*, 317. On Condorcet's response to Maupeou's coup d'état, see Elizabeth and Robert Badinter, *Condorcet: un intellectuel en politique*, 76–88.

43. Since Colbert had first received the office, the Comptroller-General had grown in importance. By the eighteenth century the office was associated with practically every other branch of government. In an era when the costs of administration and warfare had risen exponentially but the ability of the tax base to pay had declined, the necessity of improving the finances gave the Comptroller-General unusual power—paradoxically making the office itself highly insecure.

44. See Voltaire to Lalande, December 19, 1774, *Oeuvres Complètes de Voltaire*, 49: 171. "Je ne m'étonne point que des fripons engraissés de notre sang se déclaire contre M. Turgot, qui veut le conserver dans nos veins; et losqu'on nous saigne, ce soit pour l'Etat, et non pour des financiers. M. Turgot est d'ailleurs le protecteur de tous les arts, et il l'est en connaissance de cause. C'est un esprit supérieur et une très-belle âme. Malheur à la France s'il quittait son poste!" See also Voltaire to Turgot, July 28, 1774, 49, 42; Voltaire to d'Argental, September 23, 1774, 49: 84–85; Voltaire to d'Alembert, September 28, 1774, 49: 89–90; Condorcet to Voltaire, July 22, 1774, *Oeuvres de Condorcet*, 1: 36–38.

45. Information on Turgot's life from Douglas Dakin, *Turgot and the Ancien Régime in France*. Although first published in 1939, this work is still one of the most complete and useful biographies of Turgot in English.

46. Robert-Jacques Turgot, *Oeuvres de Turgot*, 1. See in particular documents from the years 1748 (pp. 102–56) and 1750 (pp. 156–238), and the *Tableau philosophique des progrès sucessifs de l'esprit humain, discours prononcé en latin dans les écoles de Sorbonne* (delivered December 11, 1750), 214–34.

47. From Du Pont de Nemours, *Mémoires sur la vie et les ouvrages de M. Turgot* (1782), cited and translated in Dakin, *Turgot and the Ancien Régime*, 11.

48. Turgot, "Fondation," *Encyclopédie*, 7. "L'utilité publique est la loi suprême, et ne doit être balancée ni par un respect supersitieux pour ce qu'on appelle l'*intention des fondateurs*, comme si des particuliers ignorants et bornés avaient eu droit d'enchaîner à leur volontés capricieuses les générations qui n'étaient point encore; ni par la crainte de blesser des droits prétendus de certains corps, comme si les corps particuliers avaient quelques droits vis-à-vis de l'Etat." "Si tous les hommes qui ont vécu avaient eu un tombeau, il aurait bien fallu, pour trouver des terres à cultiver, renverser ces monuments stériles et remuer les cendres des morts pour nourrir les vivants."

49. See Diderot to Turgot, January 21, 1759, *Oeuvres de Turgot*, 1: 594. Diderot suggested articles on *humidité*, *idée*, *idéalisme*, *intérêt de l'argent*, *impôt*, *immatérialisme*, *inspecteur*, and *intendant de province*. Strangely, Turgot scratched out *impôt* and *intendant de province* on the letter, topics in which he might reasonably be assumed to have had great interest.

50. Dakin, *Turgot and the Ancien Régime*, 16.

51. See, for example, d'Alembert's letters to Voltaire, January 11, 20, and 28, and February 8, 15, 25, *Oeuvres de d'Alembert*, 1: 51–60. As d'Alembert noted, there was real reason to fear unjust persecution: "A l'egard de l'*Encyclopédie*, quand vous me pressez de la reprendre, vour ignorez la position où nous sommes, et le déchaînement de l'autorité contre nous. Des brochures et des libelles ne sont rien en eux-mêmes; mais des libelles protégés, autorisés, commandés même par ceux qui ont l'autorité en main, sont quelque chose, surtout quand ces libelles vomissent contre nous les personnalitiés les plus odieuses et les plus infâme." Nevertheless, after a pep talk from Voltaire, he admitted the project was crucial: "Oui, sans doute, mon cher maître, l'*Encyclopédie* est devenue un ouvrage nécessaire, et se perfectionne à mesure qu'elle avance; mais il est devenu impossible de l'achever dans le maudit pays où nous sommes."

52. D'Alembert to Voltaire, September 22, 1760, *Oeuvres de d'Alembert*, 1: 73.

53. Voltaire to d'Alembert, November 17, 1760, *Oeuvres Complètes de Voltaire*, 41: 64–65. "Mon cher maître, mon digne philosophe, je suis encore tout plein de Turgot. Je ne savais pas qu'il eût fait l'article *Existence*; il vaut encore mieux que son article. Je n'ai guère vu d'homme plus aimable ni plus instruit; et, ce qui et assez rare chez nos métaphysiciens, il a le goût le plus

fin et le plus sûr. Si vous avez plusieurs sages de cette espèce dans votre secte, je tremble pour l'*infâme*."

54. Voltaire to Turgot, September 2, 1761, *Correspondance de Voltaire*, edited by Theodore Besterman, 6: 557. "Je crois même qu'il n'y a qu'un intendant qui puisse être utile. Ne peut-il pas faire réparer les chemins de traverse qui sont absolument abandonnés dans le royaume? Ne peut-il pas faire défricher des terres, dessécher des marais, encourages des manufacturers." Voltaire goes on: "Les parlements ne peuvent faire que rendre justice, et la justice est souvent un grand mal; ils peuvent aussi faire des remonstrances, et ce n'est pas toujours un bien; car elles servent à décourager la nation, et encouragent nos ennemis, il n'y a que celle du parlement de Douai qui me plaisent. Vous serez peut-être un jour contrôleur général, mais alors je serai mort."

55. Turgot pronounced his attachment to Le Mercier's and Quesnay's political ideal of *despotisme légal* in a letter to Du Pont de Nemours (the author of *Physiocracy*, 1768), November 22, 1767, *Oeuvres de Turgot*, 2: 677. Despite this, scholars have pointed out, rightly, that Turgot was never a doctrinaire or dogmatic adherent of physiocratic tenets. See Gerald J. Cavanaugh, "Turgot: The Rejection of Enlightened Despotism," *French Historical Studies* 6, no. 1 (1969): 31–58; Shovlin, *Political Economy of Virtue*, 108.

56. Turgot's reforms detailed in Dakin, *Turgot and the Ancien Régime*, chapters V–VIII, 63–117.

57. Dakin, *Turgot and the Ancien Régime*, 126.

58. Turgot to Trudaine, December 14, 1762, *Oeuvres de Turgot*, 2: 221.

59. *Circulaire aux officiers de police des villes*, February 15, 1766, *Oeuvres de Turgot*, 2: 469–77. Turgot writes, "Les moyens d'y parvenir sont de répondre avec douceur et en détail aux plaintes populaires que vous entendez chaque jour, de faire parler le langage de la raison plus que celui de l'autorité, d'engager les curés, les gentilshommes, toutes les persons qui, par leur état et leurs lumières, sont à portée d'influer sur la façon de penser du peuple, à lire l'ouvrage de M. Le Trosne, dont je vous addresse dans cette vue plusieurs exemplaires."

60. *Circulaire aux officiers* in *Oeuvres de Turgot*, 2: 470. "Il faut donc que le transport et le magasinage des grains soient entièrement libres, car, si les habitants d'une ville particulière s'arrogent le droit d'empêcher qu'on ne transporte les grains ailleurs, les autres villes croiront avoir le même droit, et les lieux où la disette est la plus grande, n'étant plus secourus par les autres, seront condamnés à mourir de faim."

61. Turgot to du Pont de Nemours, December 20, 1768, *Oeuvres de Turgot*, 3: 23. "[J]e detest la barbarie des *boeufs-tigres* et que je méprise la poltronnerie de ceux qui se laissent donner des coups de pied et des coups de corne par ces animaux aussi stupide que féroces."

62. Turgot to Terray, February 27, 1770, *Oeuvres de Turgot*, 3: 135. "Il y a des circonstances où la sagesse des administrateurs en chef est forcée de laisser une entière liberté sur le détail de l'exécution à ceux qui en sont chargés et qui résident sur les lieux."

63. See *Oeuvres de Turgot*, 4: 84–94, for various documents, letters, and memoirs concerning Turgot's projects for the navy.

64. "Lettre au Roi en prenant possession de la place de Contrôleur général," *Oeuvres de Turgot*, 4: 109–14. "Pour remplir ces trois points, il n'y a qu'un moyen. C'est de réduire la dépense au-dessous de la recette, et assez au-dessous pour pouvoir économiser chaque année une vingtaine de millions, afin de rembourser les dettes anciennes. Sans cela, le premier coup de canon forcerait l'Etat à la banqueroute. . . . J'ai senti tout le prix de la confiance dont Elle m'honnore; j'ai senti qu'elle me confiait le bonheur de ses Peuples; et, s'il m'est permis de le dire, le soin de faire aimer sa personne et son autorité. Mais en même temps, j'ai senti tout le danger auquel je m'exposais. J'ai prévu que je serais seul à combattre contre les abus de tout genre; contre les efforts de ceux qui gagnent à ces abus; contre la foule des préjugés qui s'opposent à toute réforme, et qui sont un moyen si puissant dans les mains des gens intéressés à éterniser le désordre. . . . Je serai craint, haï même, de la plus grande partie de la Cour, de tout ce qui sollicite des grâces. On m'imputera tous les refus; on me peindra comme un homme dur, parce que j'aurai représenté à V.M. qu'Elle ne doit par enricher même ceux qu'Elle aime aux dépens de la subsistance de son Peuple."

65. Unfortunately, Turgot did little to lobby for mass popular support, although he did scheme to win the support of the tax base for his policies by proposing to bypass the parlements

and establish more representative provincial assemblies to administer taxes more fairly and rationally. In preparation for this plan he ordered Du Pont de Nemours to compose the *Mémoires sur les municipalités* in August 1775. See Munro Price, *Preserving the Monarchy*, 45–48.

66. On this point, see Baker, "Science and Politics," 156–58.

67. Condorcet to Turgot, [August] 1775, *Correspondance inédite de Condorcet et de Turgot 1770–1779*, ed. Charles Henry, 234–35.

68. Badinter and Badinter, *Condorcet*, 128–29.

69. For example, Messier's expenses were covered and authorized by the government. See Condorcet to Turgot, [August] 1775, *Correspondance inédite de Condorcet et de Turgot*, 235. "Je vous envoie l'instruction pour M. Messier, hôtel de Cluny, rue des Mathurins. Il faut y ajouter une lettre pour l'autoriser à voyager et lui assurer le remboursement de ses frais d'instruments et de voyages. D'ailleurs il faut en prévenir M. de Sartines, dont il dépend comme astronome de la marine."

70. See *Oeuvres de Turgot*, 4: 648–51, and Condorcet to Turgot, *Correspondance inédite de Condorcet et de Turgot*, 252–66.

71. *Oeuvres de Turgot*, 4: 244–57, documents on "l'Épizootie."

72. *Oeuvres de Turgot*, 4: 366–67. Lavoisier also claimed that it was to French gunpowder that North America owed its independence from Britain.

73. See *Oeuvres de Turgot*, 4: 232, "Les Parlements: Edits de rétablissement."

74. Conspiracy theories abounded about the grain, including that de Conti wanted to use the disturbances to pull off a coup d'état. See Price, *Preserving the Monarchy*, 36.

75. For details on the Flour War, see Cynthia Bouton, *The Flour War: Gender, Class, and Community in Late Ancien Régime French Society*, 81–119, and Dakin, *Turgot and the Ancien Régime*, 180–92.

76. *Oeuvres de Turgot*, 4: 450. "Sans croire qu'il y ait à tout cela une première et unique cause, un complot formé et dirigé à un seul but, on ne peut se dissimuler que ce premier mouvement une fois donné a été soigneusement entretenu. . . . On a surtout été étrangement surpris de la facilité avec laquelle Paris a été pillié."

77. "Lettre d'un laboureur de Picardie à M. Necker, auteur prohibitif à Paris," *Oeuvres de Condorcet*, 11: 1–34; and "Réflexions sur le Commerce des blés," ibid., 11: 101–252.

78. The Suards in particular, who earned a pension from Necker, as well as Julie de l'Espinasse, a long-time friend of Necker. See Badinter and Badinter, *Condorcet*, 143–45.

79. Voltaire to Condorcet, August 7, 1775, *Oeuvres de Condorcet*, 1: 83–84. "Ah! la bonne chose, la raisonnable chose, et même la jolie chose que la lettre au prohibitif! Cela doit ramener tous les esprits, pour peu qu'il y ait encore dans Paris de bon sens et de bon goût."

80. Letters from Louis XVI to Turgot, May 5 and 6, 1775, *Oeuvres de Turgot*, 4: 428–29. Documents on the Grain War: 415–34.

81. For example, in those same letters of May 5 and 6, Louis XVI confessed to Turgot he sometimes drew a blank or froze while giving speeches or orders, and he admitted he was often embarrassed when speaking before groups. On the personal relationship between Louis XVI and Turgot, see John Hardman, *Louis XVI*, chapter 4.

82. "Mémoire au Roi," January 1776, *Oeuvres de Turgot*, 5: 148–62. "La suppression des corvées une fois decidée, il est également impossible de ne pas y supplées par une imposition en argent, car il faut bien que les chemins se fassent.

V.M. reconnaît la justice de charger de cette imposition les propriétaires des terres: c'est donc sur eux qu'il faut l'établir.

Cette imposition est susceptible de difficultés; mais, quand une chose est reconnue juste, quand elle est d'une nécessité absolue, il ne faut pas s'arreter à cause des difficultés: il faut les vaincre.

La première de ces difficultés consiste dans la répugnance qu'ont en général les privilégiés à se soumettre à une charge, nouvelle pour eux, que les taillables ont jusqu'ici supportés seuls.

Tous ceux qui ont à délibérer sur l'enregistrement de la loi sont privilégiés, et l'on ne peut pas se flatter qu'ils soient tous au-dessus de cet intérêt personnel, qui n'est cepandant pas fort bien entendu. Il est vraisemblable que ce motif influera secrètement sur une grande partie des objections qui seront faites. Il n'y aura même pas lieu d'être surpris que plusieurs avouent publiquement ce motif, ni même qu'ils trouvent des raisons savantes et spécieuses pour le colorer. La solution de cette difficulté est dans la justice de V.M. et dans sa volonté ferme de faire exécuter ce qu'elle lui a dicté."

83. See "Sur l'Abolition des Corvées," *Oeuvres de Condorcet*, 11: 89–97.

84. "Observations du Garde des Sceaux (du Miromesnil) et Réponses de Turgot," *Oeuvres de Turgot*, 5: 163–200. Turgot, 183: "Les dépenses du gouvernement ayant pour objet l'intérêt de tous, tous doivent y contribuer; et plus on jouit des avantages de la société, plus que on doit se tenir honoré d'en partager les charges. Il est difficile que, sous ce point de vue, le privilège pécuniaire de la noblesse paraisse juste."

85. See Dakin, *Turgot and the Ancien Régime*, 249–50.

86. L'abbé de Veri, *Journal*, 1: 448. See Price, *Preserving the Monarchy*, 49.

87. This letter and others from Turgot's last days in office have been lost and are reported in Veri, *Journal*, 1: 450–57. See Dakin, *Turgot and the Ancien Régime*, 261.

88. "Projet de letter de Turgot à Maurepas," *Oeuvres de Turgot*, 5: 466–67. "Je veux donc essayer de vous rappeler à vous-même, à ce que vous avez mille fois pensé et dit, à ce que vous devez au public, au Roi, à votre propre reputation, d'opposer votre conscience à la passion, qui, permettez-moi de vous dire, vous déguise jusqu'à votre propre intérêt."

89. Voltaire to Condorcet, September 7, 1776, *Oeuvres Complètes de Voltaire*, 50: 81. "Je suis dans une apertume continuelle depuis qu'on nous a ôté le protecteur du peuple et celui de ma province. Depuis ce jour fatal, je n'ai suivi aucune affaire, je n'ai rien demandé à personne, et j'ai attendu patiemment qu'on nous égorgeât." See also Voltaire to d'Alembert, June 10, l, 30–31; to La Harpe, June 10, ibid., 50: 32. "Je vous prierai surtout de me dire où est l'intrepide philosophe M. de Condorcet. Est-il à Paris? N'est-il occupé à consoler M. d'Alembert? Ni eux ni moi ne nous consolerons jamais d'avoir vu naître et périr l'âge d'or que M. Turgot nous préparait." And to Dupont, June 15, ibid., 50, 43. "M. Turgot, le père du peuple m'encourageait . . . la destitution de ce grand homme m'ecrasse, et je vais mourir en le regrettant."

90. Condorcet to Voltaire, 1776, *Oeuvres de Condorcet*, 1: 113.

91. Condorcet to M. Suard, May 1776, *Correspondance inédite de Condorcet et Mme Suard, M. Suard et Garat: 1771–1791*, 176–77. "Il est triste de ne plus travailler que pour son plaisir . . . lorsqu'on a eu, pendant quelque temps, l'espérance de travailler pour le bien des hommes."

92. Condorcet to Voltaire, 1776, *Oeuvres de Condorcet*, 1: 113. "Je ne vous ai point écrit, mon cher et illustre maître, depuis l'événement fatal qui a ôté à tous les honnêtes gens l'espérance et le courage. J'ai attendu que ma colère fût un peu passée, et qu'il ne me restât plus que de l'affliction. Cet événement a changé pour moi toute la nature. Je n'ai plus le même plaisir à regarder ces belles compagnes où il eût fait naître le bonheur. Voilà où nous sommes tombés, mon cher et illustre maître, et de bien haut! Adieu, mon cher et illustre maître; nous avons fait un beau rêve, mais il a été trop court. Je vais me remettre à la géométrie et à la philosophie. Il est bien froid de ne plus travailler que pour la gloriole, quand on s'est flatté pendant quelque temps de travailler pour le bien public."

Chapter Four

Struggle and Radicalization on the Eve of the Revolution: Condorcet and the Transformation of the Idea of the Sciences

When the marquis de Condorcet formally succeeded Grandjean de Fouchy as *secrétaire perpétuel* of the Royal Academy of Sciences in August 1776, the world had become, at least for Voltaire's faction of the *parti philosophique*, a very different place. The disgrace and dismissal of Turgot from the ministry the previous May filled many of the philosophes with frustration and despair. For Condorcet, who was the youngest of the group associated with Voltaire, the failure of 1776 marked a decisive moment: the fall of Turgot ultimately precipitated Condorcet's abandonment of the absolute monarchy as the agent of reform and his formulation of a republican, and even proto-democratic, ideology as a political alternative.[1]

Some have argued that the fall of Turgot made the Revolution of 1789 inevitable.[2] Historical inevitability is, of course, difficult to gauge, but it seems clear enough that after the fall of Turgot, Condorcet began to produce writing that had more radical, and even revolutionary, implications. The Voltairean faction, whose intellectual leadership was now passing to Condorcet, experienced with the fall of Turgot a radical disjuncture between their principles and the world—for the world seemed insistently to refuse to adhere to these philosophes' moral idea of sciences. In such an unsatisfactory situation, either the idea or the world must give. The political consciousness that formed among Voltaire's associates over the course of the first half of the eighteenth century was ill disposed to accommodation with an unacceptable world, for the rhetoric by which this political consciousness was articulated posited a Manichean struggle between the forces of good and evil that did not allow much room for conciliation. To Condorcet, in particular, there could be no acceptance of Turgot's defeat as the end of the struggle, as a time to come to terms with the enemy. Instead, Turgot's defeat prompted Condorcet to transform the idea of the sciences as morally enlightening—an idea that had already come to contain radical implications with Fontenelle's lionizing of savants—into something with overtly revolutionary meaning in which savants would seize political leadership and reform the world—by force, if necessary. The self-restraint and intellectual modesty—the antidogmatism—that characterized Samuel Sorbière's idea of the sciences in the mid-seventeenth century is hard to find in Condorcet. Although Sorbière would have recognized Condorcet's goals for reform as emerging from the same Neostoic source, Sorbière would hardly have recognized the hubris of this faction of the philosophical party.

We can trace the contours of Condorcet's ideological transformation by examining the development of his and his ideological associates' rhetoric over the second half of the eighteenth century. A rhetoric of struggle emerged among the philosophes associated with Voltaire in the mid-eighteenth century. This rhetoric was the means by which these philosophes expressed, in a sort of feedback loop, both their frustration over and their understanding of the disjuncture between the perceived irrationality of the sociopolitical world and the rational scientific ideals that disposed them to interpret the world as they did and which, thus, informed their moral and political discourse. Through rhetoric they defined an ideology—a political position on which to act. Voltaire remains an important part of this story for promoting the apocalyptic language that articulated so clearly the experience of disjuncture. Here I am most interested in Voltaire's protégé, Condorcet, however, for only Condorcet had the opportunity to become a revolutionary—an opportunity he took. He was one of the first, possibly *the* first, of the now famous philosophes to declare himself publicly in favor of deposing the king and establishing a republic in July 1791, just after the royal family's abortive flight to Varennes.[3] In the process of becoming a revolutionary, Condorcet transformed the idea of the sciences as morally enlightening into a revolutionary ideology that clarified his motives, thus enabling him to engage coherently in revolutionary actions when the opportunity arose.

In the previous chapters, I examined the articulation of a moral-political discourse—the idea of the sciences—emerging after and in partial response to the religious wars and the Frondes, which located the source of civic and personal virtue in the pursuit of the natural sciences.[4] We see this discourse in the speeches and writings of savants such as Samuel Sorbière in the 1660s, but especially after the 1680s with Bernard de Fontenelle, who argued that savants achieve virtue through their study of the sciences. By the mid-eighteenth century, through multiple political, economic, and military crises, this idea of the sciences had become a primary component of what are often now viewed as quintessentially enlightenment ideals, deployed by Voltaire to suggest, as he did in the *Le Siècle de Louis XIV*, that the sciences had brought about a "revolution in the human spirit" that was rapidly ridding the world of ignorance, superstition, and error. Far from politically revolutionary, the "philosophical spirit," Voltaire argued, "contributes much to legitimize the rights of sovereigns."[5] Voltaire's words were not, of course, objective observations of historical fact, but a rhetoric through which he staked out an ideological position and sought to convince his readers that the political, military, financial, and even moral crises of Louis XV's reign could be resolved if the state were reformed to conform to rational scientific ideals, to exhibit the "philosophic spirit."[6] In the midst of the conflict between the monarchy and the Assembly of the Clergy over the proposed taxation of church property in summer 1750, for example, Voltaire insisted that "there is in this world not a single example of philosophers who have opposed the laws of the prince; [but] there is not a single century in which superstition and enthusiasm have not caused troubles that horrify us." "The best thing that can happen to men is for the prince to be a philosopher," he continued. "The philosophic prince knows that the more progress reason makes in his state, the less damage will be done by disputes, theological quarrels, enthusiasm, and superstition; hence he will encourage the progress of reason."[7]

Voltaire's younger associates, particularly d'Alembert, Turgot, and Condorcet, joined him in voicing and then acting upon his ideas in the 1760s and 1770s. As the monarchy increasingly failed to live up to the Voltairean faction's goals, however, and as the perceived gap between rational, scientific discourse and the sociopolitical order grew wider, these philosophes expressed their perception of sociopolitical irrationality as frustration and disillusionment. The rhetoric to emerge from their ultimate despair portrayed the philosophic cause as a momentous and even biblical struggle against darkness. Voltaire, a relentless cheerleader for the cause, defined the nature of the battle as being against intransigent parlements and intolerant clergy—not, as historians have long shown, a necessarily fair or accurate assessment, but it simplified matters to clarify sides. In 1759, for example, after the Parlement of Paris condemned the *Encyclopédie*, Voltaire pressed d'Alembert not to give up on the project, no matter the persecution he must face—although Voltaire did agree that prudence counseled moving publication abroad, an idea that Diderot rejected as cowardly.[8] In November of the following year, after granting an audience to the young Turgot at Ferney, Voltaire wrote encouragingly to d'Alembert, "If you have several wise men of this sort in your party, then I tremble for the villainous [*l'infâme*]."[9] In 1761, when Turgot's willpower flagged, Voltaire refused to allow him to abandon his public service: "I believe that an Intendent is the only person who can be useful," he insisted.[10] "The parlements can only render justice," he went on, "and this justice is often a great evil. They can also make remonstrations, and this is not always good because these remonstrations discourage the nation and encourage our enemies."[11]

Such rhetoric was self-reinforcing and grew more intense over time. By defining an ideological position with goals to achieve and enemies to vanquish, this rhetoric also inspired action that further reinforced the rhetorical intensification—not least by providing, and even helping to create, a more solid idea against which enemies of *philosophie* could define themselves. As Darrin McMahon has shown, *anti-philosophie* also emerged as a more well-defined ideology in the mid-eighteenth century, a reactionary mirror image of the radicalizing ideology of *philosophie*, each reinforcing the other.[12] We have already witnessed the case of Turgot as he fought the parlements during his long intendancy of Limoges (1761–1774) and his short tenure as Comptroller-General of Finances (1774–1776). Turgot's strong words and heavy-handed actions polarized sentiments both for and against the philosophic cause, and it led with apparent inexorability to a climactic battle: ultimately, the so-called Flour War of 1774 and the conflict over Turgot's infamous Six Edicts of 1776—all resulting in his fall and disgrace in May 1776. To bring him down and discredit the political goals of *philosophie*, Turgot's enemies resorted to whatever means necessary, including libel, assaults, petty crimes, lies, and forged correspondence purporting to confirm treasonous plots between Turgot and his friends. For its part, the Voltairean faction did not scruple to engage in polemic, slander, and force to defend its goals (see chapter 3).

Condorcet gave himself wholeheartedly to the intensification of the rhetoric of struggle. For Condorcet, the fall of Turgot revealed a chasm between the world and his ideas, leading him to conclude that the monarchy could not be the agent of reform. Voltaire's rhetoric reinforced such perceptions. His June 1776 correspondence likens Turgot's fall to the end of a golden age: "Where is the intrepid philosophe M. de Condorcet?" he wrote to Jean-François de la Harpe. "Is he in Paris? Is he occupied with consoling M. d'Alembert? Neither they nor I will ever console us for having seen born and die the golden age M. Turgot was preparing for us." And to Du Pont, "The fall of this great man crushes me."[13] In September 1776, Voltaire wrote to Condorcet, "Since that fatal day, I have not followed anything, I have not requested anything of anybody, and I am waiting patiently for someone to cut our throats."[14]

Because the rhetoric of struggle (as a structuring structure) disposed the Voltairean faction to interpret the world in such stark terms, the failure of Turgot's reforms indicated to those philosophes most attached to the idea of the sciences that the monarchy could not accomplish desired rationalizing reforms.[15] The continued failure of later attempts in even more dire financial and political circumstances by Comptrollers-General Calonne and Brienne to revive all or part of Turgot's plan only strengthened perceptions that the absolute monarchy had failed and could not hope ever to succeed.

But if not the absolute monarchy, then what? For all his adult life, Voltaire, for one, had championed absolute monarchy as the only practical alternative for France. Fontenelle before him had believed the same, as had Sorbière before him, and the other *libertins érudits* and Neostoics dating back to the end of the sixteenth century. Turgot's fall was therefore a sobering realization of the failure of reason. Voltaire mourned not only the disgrace of a friend and ally, but also, as he put it, the end of a potential golden age. Other philosophes reacted with similar despair. D'Alembert, distracted yet further by the death of his beloved Julie de Lespinasse (also in May 1776), withdrew almost entirely from public intellectual life. Of Turgot's ministry, he had once noted that "if good does not come of this, one must conclude that the good is impossible."[16] His conclusions seemed to him unfortunately correct. Turgot himself retired immediately from all political and government involvement and concerned himself until his death only with astronomy and what Fontenelle had called the *plaisirs simples*.

The older generation was perhaps ill prepared to contemplate a radical modification of long-held political views. Some sophisticated thinkers, such as Diderot, had always been uneasy about the despotic tendencies inherent in absolute monarchy, and he reacted to Turgot's dismissal with fewer histrionics.[17] Voltaire and his lieutenants in Paris, however, had been the vociferous enemies of the parlements for too long to accept a political order in which the parlements appeared to assume the upper hand, and they made little of the vague, generally ignored calls for an Estates-General heard occasionally after the Maupeou coup d'état of 1771.[18] No new ideas came from an older generation—a generation that was passing away: Voltaire died in 1778, Turgot in 1781, d'Alembert in 1783, and then Diderot, too, in 1784.

But Voltaire and d'Alembert had groomed Condorcet in the late 1760s and 1770s to take up the ideological leadership of their philosophic faction. Under the rays of Voltaire's heated rhetoric, the morally enlightening idea of the sciences became for Condorcet a fervent impetus for political action.[19] By 1776, when he became perpetual secretary of the Academy of Sciences, Condorcet had developed a distinctly politicized view of the sciences that was born during his experiences in the atmosphere of crisis in the 1770s and in the heady days of Turgot's ministry. Still young (he was thirty-three in 1776) and impassioned by his work for Turgot as inspector of the mint and inspector-general of navigation, Condorcet revealed a great eagerness to assert a leading political role for savants and philosophes in the state—if not as servants of the monarchy, then as public leaders themselves.

This is not to imply that Condorcet had conceived in 1776 of an enlightened republic, or a sort of "aristocracy of the learned."[20] The overtly politicized idea of the sciences that he had started developing in the early 1770s, however—formed in the dialogic of ideas and experience—laid the foundation for the development of more radical political views. The rest of this chapter, therefore, examines the maturation of Condorcet's ideas on science and philosophy in the 1770s. The following, final chapter will explore the evolution of these ideas into a coherent political alternative to absolute monarchy that was consistent with the goals of philosophes who had for so long seen the monarchy as the only viable political choice: an enlightened republic based on universal education in the sciences.

Condorcet's regret at having lost an opportunity to act for the common good characterized a zeal for public action central to his developing moral and political views. Fortified by his faith in the sciences and influenced by the example of Turgot, Condorcet's commitment to public action rivaled Voltaire's by the mid-1770s (although it certainly lacked Voltaire's subtlety and wit). The cultivation of this perhaps naturally latent fervor was due in no small part to Voltaire, as Voltaire's reply to Condorcet's May 1776 letter (see chapter 3) begins to indicate: "As for you," Voltaire wrote back in July, "I believe you now ensconced in geometry. I forgive you for it if you make new discoveries in mathematics, as did Sir Isaac and Captain Halley."[21] Like the *libertins érudits* from whom Voltaire had inherited many of his own political and moral views, he would not countenance a withdrawal from the world into a realm of philosophical tranquility. Voltaire's Stoicism was still that of du Vair or Charron. Happiness and tranquility might be the goal of life, but activity of no practical use to humanity, especially in a period of crisis, lacked virtue. Even after the distressing events of May 1776, Voltaire could only forgive Condorcet's threat to withdraw from the public sphere if he were to engage in useful scientific innovation or discovery.

Historians have been tempted to explain Condorcet's quasi-religious and remarkably humorless fanaticism for the philosophic cause through a psychological analysis of traumatic childhood experiences.[22] Such an explanation may have merit so long as it is tempered by recognition of the role Voltaire and his rhetoric of struggle later played in Condorcet's intellectual maturation. As the intellectual successor of Fontenelle and the leading champion of the political and social application of scientific values, Voltaire was perhaps the most important intellectual influence on Condorcet. From the mid-1770s, Voltaire clearly intended the young man to be his intellectual and even ideological successor—and not just metaphorically: Voltaire hoped Condorcet would take his place in the *Académie Française*.[23]

From the ancient but impoverished nobility of Picardy, Marie Jean-Antoine-Nicolas Caritat, marquis de Condorcet (1743–1794), had opposed the wishes of his family by becoming a mathematician. Although the academicians d'Alembert, Grandjean de Fouchy, and Étienne Bézout, who judged his impressive bachelor's thesis at the Collège de Navarre in 1759, had "saluted him as one of their future colleagues," his family resisted his career choice. His uncle, the bishop of Lisieux, had intended him for the cavalry, but Condorcet refused to take up a military career. His decision to move to Paris in 1762 to pursue further studies caused persistent rancor in the family. From then until 1776 he lived on a small pension from his mother, renting a room and eating in.[24] His work, particularly *Du calcul intégral* (1765), nevertheless attracted the attention of the Parisian mathematicians, who offered him an unsalaried position as *géomètre associé* in the academy in 1768. Still hoping to dissuade him from an ignoble career as a savant, Condorcet's family initially blocked his acceptance of the position, but, seeing such moves were ultimately useless, they relented in 1769. In April d'Alembert wrote to Lagrange, apprising him of the news: "We have finally received M. de Condorcet, his family having judged it no longer appropriate to obstruct his entrance into the academy, although many of our gentlemen believe that the title and profession of savant derogates nobility."[25]

We might interpret Condorcet's resistance to the military, the traditional vocation for members of the old nobility, as something of a response to early childhood experience: Condorcet's father was killed during training maneuvers with the cavalry in 1743, and his pious widow was so traumatized by the event that she became excessively protective of her only son, Jean-Antoine. The story has it that she dedicated him to the Virgin, kept him in skirts, and did not allow him to engage in normal male childhood activities. Whether this had some effect on his attitudes later in life is speculation, of course (and dressing a male toddler as a girl was hardly unusual in the eighteenth century), as is the supposition that later negative experiences at his Jesuit school in Rheims led him to reject superstition and religion and ardently to embrace mathematics—the language of demonstrable truth and reason—as the basis for his social and moral ideals. Modern biographers of Condorcet assert that, although there is no evidence of any mistreatment or abuse, Condorcet's later hatred for the Jesuits and religion in general, and his contempt for their style of education, points to something more than mere dispassionate disapproval.[26] In citing the lack of evidence for abuse, they imply it occurred, of course, but we should note that most of the anti-clerical and anti-Jesuit writing we have from Condorcet is from no earlier than the 1770s, after he had met Voltaire—and, particularly, after the execution of the Chevalier La Barre in 1766 for heresy.[27]

In 1766 the Parlement of Paris effectively declared war on *philosophie* with the torture and execution of the Chevalier La Barre for, according to Voltaire's version of the case, little more than refusing to doff his hat before a processional and vandalizing a wooden crucifix with his cane (the burning of La Barre's corpse along with a copy of Voltaire's *Dictionnaire philosophique* only added insult to injury).[28] Disgusted by the display of intolerance and cruelty, d'Alembert later called the French justice system a "masterpiece of atrocity and stupidity."[29] Voltaire, with yet another reason to detest the parlements, gave the magistrate in charge of La Barre's sentence, Pasquier, the nickname "boeuf-tigre"—an insult that thereafter stuck with the *parlementaires*, implying stupid ferocity. Voltaire also sought to protect La Barre's "accomplice," Gaillard Étallonde, who had fled France rather than face trail.[30] By 1774, with Turgot in power, Condorcet became the chief Parisian advocate of Voltaire's campaign to have La Barre's conviction posthumously overturned.[31] The attempt to reform the French legal system failed with the fall of Turgot, but it led to the further intensification of the rhetoric of struggle.[32] "The agony of superstition will be long," Condorcet wrote to Voltaire in March 1774, "and it sometimes has intervals of strength that horrify!"[33]

Since taking up residence in Paris in 1762, Condorcet had moved among the philosophic community. D'Alembert introduced him into the salon of Julie de Lespinasse, where he met other philosophes such as Turgot, Melchior Grimm, and Baron d'Holbach, as well as Amélie Suard, the sister of C. J. Panckoucke, publisher of the *Encyclopédie*. Through d'Alembert, Condorcet also became one of the last contributors to the *Encyclopédie* himself, editing the articles on mathematics in the new editions of 1776 and 1781. Perhaps most important to his future ideological development, Condorcet accompanied d'Alembert in 1770 on a voyage to visit Voltaire at Ferney, some ten years after d'Alembert had sent Turgot on a similar visit. As Condorcet's first biographer, François Arago, pithily points out, Voltaire was the Dalai Lama of the intellectual world, and his influence on young philosophes must not be underestimated.[34] Whatever the content of the Voltaire-d'Alembert-Condorcet entente of 1770—we, unfortunately, have no account of how they passed their time together—the result was a concerted campaign to obtain the secretariat of the Academy of Sciences for Condorcet and that of the Académie Française for d'Alembert, and thus assume leadership of two of the most influential intellectual positions in France. Voltaire surely communicated to Condorcet some of his apocalyptic (even messianic) fervor so as to *ecrasez l'infâme*, as well as his opinions on the duty of the *esprits supérieurs* to lead the way into battle.

In Condorcet, even more than in Voltaire, the rhetoric of struggle that characterized the discourse of *philosophie* would take on the Manichean aspect of a millennial battle between the forces of light (reason) and darkness (ignorance). Such dualism turns up most overtly in Condorcet's final, posthumously published work, the *Esquisse d'un tableau historique des progrès de l'esprit humain* (1794), but it originated much earlier. In the 1780s, when he began to articulate publicly an alternative to absolutist monarchy that would require the political leadership of philosophes and savants, he sought to prepare them for the battle by identifying clearly (if not originally) the enemy, relying on the same themes from Neostoic thought we have seen with Sorbière and Fontenelle: "We now know that the true enemy of mankind is ignorance; that it produces both the passions that trouble the order of the world, and the weakness that makes these passions dangerous."[35] As early as 1771, however, just after his visit with Voltaire, Condorcet had begun to think of the goals of the *parti philosophique* in terms of interminable struggle. In a letter to Turgot in January, he quoted a line from the *Mercure*: "The most sure means of making men better and more happy is to enlighten them, and the esteem of those who enlighten nations is alone able to ease the pains of government." Acerbically he added, "See there what some princes have written—but these princes, educated by philosophers, still have no idea of the noble, enlightened, and honest politics of our great men of state." By "great men of state," he meant, of course, men such as Turgot who would "ease the pains of government" and enlighten the people (regardless of whether they wanted such enlightenment). The ignorant, enslaved to their base passions, had, after all, no ability to judge their proper needs. The duty of the educated was to judge for them. "I would like to see you again," he added to Turgot, who still served as intendant in Limoges, "and discuss with you the misfortunes of reason and humanity, and our hopes, if any still remain, and the consolations that we must await with time, that will put men and things in their places."[36]

The enemies of *philosophie* were not limited to the clergy and the parlements. Anyone at all—even other savants—who stood in the way of the Voltairean faction's ideological goals could earn excoriation, as we see during d'Alembert's and Condorcet's heated and contested campaigns to become *secrétaires perpétuels* of the French Academy and the Royal Academy of Sciences, respectively. Some academicians regarded d'Alembert's and Condorcet's political maneuverings as naked attempts to gain control of the academies for their own self-interested purposes—in particular, to protect and promote the continued publication of the *Encyclopédie*. Voltaire and his associates naturally interpreted any opposition to their plans as opposition to the philosophic cause—that is, to seize the reins of power and so join power and reason in perfect unison. When d'Alembert was finally elected secretary of the Académie Française in 1772, Condorcet conveyed the news to Voltaire in biblical terms: "Our friend is [now] perpetual secretary of the Académie Française. The enemies of philosophy defended themselves well, but the soldiers of Gideon will always vanquish the Midianites, knocking them aside with the force of their light."[37]

Condorcet's own struggle to become secretary of the Academy of Sciences faced much more controversy, during which he and d'Alembert revealed to the other academicians a ruthless zeal that contrasted starkly with the civility that ideally characterized the culture of the Republic of Letters. Aware that the aging Grandjean de Fouchy would soon retire from office, d'Alembert engaged in machinations and intrigue to secure the post for Condorcet before it was technically available. Suggesting that age had made

Fouchy negligent in his duties, he argued that, contrary to custom, Fouchy needed a young academician to assist and eventually succeed him.[38] Fouchy, who had been impressed with Condorcet since sitting on his bachelor's thesis committee, eventually agreed to this plan and proposed Condorcet's elevation to the post as a *suppléant en survivance* at the beginning of 1773 in the midst of fierce opposition. In the meantime, however, d'Alembert sought to prove Condorcet's suitability for the position by a demonstration of his talents. In the spring of 1772 he quietly suggested to Condorcet that he compose a eulogy of the celebrated mathematician, Alexis Fontaine des Bertins, who had died in 1771.

D'Alembert was, of course, well aware that by the 1770s the practice of eulogizing had, since Fontenelle, become central to a secretary's duty; anyone who sought the position had to demonstrate special talent for it. As an academician wrote anonymously in 1773, "The secretary must have a talent for writing, if not as well as the illustrious Fontenelle, then at least with purity, clarity, and excellence." Since the Academy owed much of its popular image to them, the eulogies in particular must honor the memory of the academicians such that their sacrifice for the common good is obvious. "We will not at all discuss here the qualities of style proper to the eulogies," continued the anonymous academician, "we will only say that they must be interesting, so that the public can hear and read them with pleasure, and understand the deeds of those great men who have died for them and for the good of the nation."[39]

Leaving d'Alembert in Paris to promote his candidacy, Condorcet returned to his hometown of Ribemont in Picardy in the summer of 1772 to devote himself to the eulogy and to visit his aging mother. Preparing an *éloge historique* for an academician of Fontaine's quality required scholarly research, however, and Condorcet was impatient to continue the philosophic struggle more actively. "You are very fortunate to have a passion for the public good and the power to satisfy it," he wrote to Turgot from Ribemont. "It is a consolation of an order superior to that of study."[40] Turgot, perhaps weary of his endless duties and longing for the philosophic life he had turned down in 1761, replied with a mild rebuke: "Whatever you might say about it, I believe the satisfaction resulting from study to be superior to that of all others. I am very convinced that one can be by it a thousand times more useful to men than by all inferior activities, where one torments oneself to do a few small goods, often without success, because one is the forced instrument of great evils. All these minor goods are temporary, whereas the enlightenment that a man of letters can bring ought, sooner or later, to destroy all the artificial evils of the human species and allow it to enjoy the goods that nature offers it."[41]

Concerned that he had been too hard on his young friend, Turgot wrote again a month later asking for news, adding that Condorcet need not fear any further "reprisals" from him. "Mademoiselle de Lespinasse tells me you are working on some literary piece," he continued, lamely. "I do not believe it would be an indiscretion to ask you what it is about."[42] Although Condorcet hitherto had kept the nature of his project secret from all but d'Alembert, he sent a copy of the eulogy to Turgot for comments. Turgot's rebuke seems to have helped Condorcet focus the eulogy effectively. In order to satisfy the Academy, a eulogy had necessarily to follow Fontenelle's lead, and Turgot's remarks had been a concise interpretation of Fontenelle's idea of the sciences as morally enlightening but well in line with Condorcet's more activist sensibilities: the enlightenment obtained through the study of the sciences helped savants achieve the virtue necessary to aid their fellows to reclaim their natural rights. "I will profit with great pleasure from your remarks on the eulogy of Fontaine," Condorcet wrote in October, after receiving Turgot's corrections. "I will certainly add to it a word on the necessary obscurity of the account I made of his principal works, and at the same time I will clarify their far-reaching utility. If you want, I will send you this addition."[43]

Even in his first eulogy Condorcet adopted an idiosyncratic style that differed from the work of his predecessors. Rather than merely attempt to copy, with less literary skill, Fontenelle's masterful, neoclassical style, Condorcet struck a less formal, more utilitarian tone. Not concerned with making the academician appear inhumanly perfect, Condorcet related considerable biographical detail, readily described the foibles and faults of his subject as well as his virtues, and did not scruple to reveal Fontaine's vain, scornful, and vitriolic personality. Condorcet even deigned to insert himself into the narrative where appropriate.[44] "One only owes to the dead what is useful to the living," he noted later, "truth and justice."[45] Philosophically, however, the eulogy was deeply indebted to Fontenelle's interpretation of the idea of the sciences. Despite a description of Fontaine's disagreeable disposition, Condorcet painted Fontaine's virtues in unmistakably Fontenellian terms: According to Condorcet, Fontaine had a passion for truth that made him indifferent to the false, ephemeral pleasure of personal glory. He was an ascetic, living solitary and preferring the simplicity of the country to the decadent pleasures of the city. "Such was this man, as original perhaps by his character as by his genius, that society had neither corrupted nor polished him, who inclined toward all the pleasures of nature, and followed them without constraint, but who never knew violent passions, which are the result of society." According to the Fontenellian ideal, Fontaine was a perfect philosopher because he had discovered through the sciences those true goods that directed his passions toward simplicity and tranquility, thus allowing him to achieve virtue (see chapter 2). "Happy while studying," wrote Condorcet, "and having in geometry and in the cultivation of the earth a sure remedy against sadness, M. de Fontaine had no need either of the services, or unfortunately for him, the friendship of anyone. He never sought to please, and disdained making himself feared." Even while frequenting the city, Fontaine knew that the immoderate pleasures it offered were temporary and false, incapable of leading to true happiness. In the last years of his life, the sciences had engendered in Fontaine such detachment and tranquility that he could accept death with perfect stoicism: "He had always regarded pain and death as the necessary result of the general laws of nature, of which it would be absurd to complain."[46]

In his first eulogy Condorcet demonstrated a clear understanding of Fontenelle's moral philosophy of the sciences: that the sciences themselves led the *esprit supérieur* to direct his passions toward moderation, simplicity, and tranquility, and thus enabled him to achieve virtue. Condorcet was not content, however, to merely copy Fontenelle's ideas. Encouraged by the warm approval of his friends regarding the eulogy, Condorcet continued to write eleven more eulogies by December that would take Fontenelle's moral philosophy to the next level and assert an active political role for the virtuous savant.[47]

Fortunately for the Academy, eleven academicians had not suddenly died. Condorcet chose to eulogize those academicians from the earliest days of the Academy, before the time of Fontenelle, for whom no perpetual secretary had ever written eulogies.[48] The eulogies were published near the end of December, and Condorcet followed Fouchy's custom of inserting a comment about Fontenelle in the preface:

> According to the Regulations of 1699, the secretary must present eulogies of those academicians who have passed away. M. de Fontenelle did this up to the year 1740. His collection of eulogies is one of those books that can be read over and over. And one never rereads it without discovering in it new charms, and without admiring his very rare talent of being clear about the most difficult matters, of saying the most common things with finesse, and the most subtle with a simplicity that renders them even more piquant. The savants who died before 1699, however, did not have eulogies, and thus Fontenelle's work lacked this history of the sciences. I have tried to supply it here.[49]

Despite the necessity to commence the publication with a paean to Fontenelle, the eulogies were not simply additions to the work Fontenelle had begun. Condorcet had his own agenda. His eulogy of Roberval was, for example, hardly a eulogy at all, but the continuation of the critique—begun with regard to Roberval perhaps by Sorbière—of petty ambition and its destructive results for the scientific enterprise, for only by working for the common good could the sciences claim their moral authority. Condorcet, therefore, relentlessly criticized Roberval for his overbearing arrogance and infamous vanity. He pointed out Roberval's failures in detail, and he explained the degree to which they galled the man, leaving him bitter and frustrated. "This unworthy vanity," he wrote, "prevented Roberval from achieving either reputation or repose. If he had studied the geometry of Descartes instead of combating it, he would have been the first among his disciples." Furthermore, Condorcet asserted, in creating an atmosphere of internecine conflict, Roberval endangered not only the progress of scientific innovation but also the attempt to secure public and royal support for scientific institutions: "Why should it be necessary that such men, made to contribute to the progress of the sciences, squander in combating a superior man the strength they ought to have employed against the resistance of nature? Why must they sacrifice so miserably to the pettiness of vanity the interests of the sciences and that of their true glory?" Roberval had studied the sciences in the hopes of achieving fame and glory among other men—a goal that Fontenelle had shown was merely a fleeting pleasure that could never lead to tranquility and virtue. Controlled by his immoderate passions, Roberval had never practiced the sciences philosophically. Condorcet concluded, "In his methods, nearly all obscured and overturned, one nevertheless cannot fail to recognize the mark of genius."[50]

In opposition to his characterization of Roberval, Condorcet invoked the image of the other, more properly motivated savants, including the virtuous abbé Picard. Working tirelessly in the 1670s to discover the circumference of the earth, Picard did not suffer from Roberval's failings, according to Condorcet: "Less concerned with his reputation than with the pleasure of giving to his country a man who would be useful to it, he did not at all fear having a rival occupied with the same research, even if it could detract from his own glory."[51]

Whereas Fontenelle focused on the personal virtue and happiness that *esprits supérieurs* acquired through the sciences, Condorcet illuminated the public duty savants' moral and intellectual superiority required of them. This duty was in no way limited to the Academy, but it extended into the political realm. In his eulogy of the Danish astronomer, Røemer, Condorcet noted that Frederick IV of Denmark had made Røemer first magistrate of Copenhagen in 1705. "Frederick IV," wrote Condorcet, "was uninfluenced by the unfounded opinion in many princely courts that savants are not capable of administration, as if the habit of seeking after the truth could not replace the habits acquired in inferior employments." Indeed savants, fortified by study and the discovery of practical knowledge, were the proper guardians of the public good, moralized Condorcet: "If, however, one takes the spirit of intrigue for that of normal affairs, and the art of fooling or oppressing men for that of governing them, one is right to believe that savants are not proper for the job. A soul that has for a long time nourished itself with the love of truth and glory is hardly able either to sense the necessity or take the habit of this mix of falsity and meanness that one decorates with the name of political adroitness."[52]

Better equipped intellectually and morally to administer society than the traditional political elite, Condorcet suggested, savants were right to take an active role in the state, to strive for a more just and rational society. The struggle between the forces of reason and ignorance on which Condorcet elaborated in his posthumous *Esquisse* (published in 1794) had already begun to take shape by the end of 1772. In his eulogy of Claude Perrault, the formation of the Académie royale des sciences itself took on a conspiratorial aspect as philosophes worked behind the scenes, pulling the strings of power, in order to "recall men of letters to the dignity of their state" and achieve their noble goals: "It was [Claude Perrault and his brothers] principally who inspired Colbert to protect the sciences in a manner so flattering for savants and so honorable for himself. The Cardinal Richelieu had seen in men of letters only men useful for his own glory. Colbert, directed by Perrault, treated them as men useful to humanity, whose glory honored their country and their century."[53]

Condorcet's friends and supporters responded to the eulogies with predictable enthusiasm. To Fontaine's eulogy, Voltaire responded with high praise: "You are certainly my master, Monsieur le Marquis, and you would have been Bernard de Fontenelle's. You are a true and eloquent philosopher. . . . You nourish and embellish a dry subject with a noble and profound moral." Later, upon receiving a copy of the published eulogies, he wrote again, "You appear to be the master of this subject, but a gentle and modest master, as a king who writes the history of his subjects."[54] Joseph-Louis Lagrange also sent congratulations: "The simple, noble, and true style in which they are written seems the only proper one for these matters and renders them infinitely superior to many others that shine only by an affected and high-flown style."[55] D'Alembert and his ally, Bossut, hoping to sway academic opinion in Condorcet's favor, spoke glowingly of the eulogies and recommended the assembly express its gratitude publicly.[56]

Grimm was not persuaded by such rhetoric, however, and noted in the *Correspondance littéraire* in February, "They say that M. de Condorcet showed as much spirit as, and better taste than, Fontenelle. His friends, in overrating and exaggerating, make a mistake and spoil everything. I would desire in general from M. de Condorcet a style a little more interesting."[57] Others in the Academy—namely, Buffon and his cadre, who had their own candidate for the secretariat (the astronomer Jean Sylvain de Bailly)—also saw through d'Alembert's machinations and complained bitterly in March when Fouchy formally asked for the confirmation of Condorcet as his co-secretary. The *parti de Buffon* assumed, with some reason, that d'Alembert would be the true power behind Condorcet's secretariat and would therefore have considerable control over the most prestigious learned societies in Paris: the *Académie Française* and the Academy of Sciences. Protesting the break of protocol that seemed to set Condorcet up to succeed Fouchy, they issued a statement summarizing their position:

> Without doubt personal interest, or even without supposing this motive, a certain natural inquietude in some minds, makes them see some sort of illusory advantage in this change [of tradition]. But for a hundred years the Academy of Sciences has worked well having but one secretary. When after 80 years of age the illustrious Fontenelle—who alone immortalized the history of the sciences, and who is worth several other men because of the universality of his knowledge and a truly philosophic vision—surrendered the burden of his functions, he was nevertheless replaced by one man alone.[58]

Whatever the merits of Buffon's fulminations—and he was hardly exempt from personal interest in the outcome—the king approved Condorcet's nomination as co-secretary on March 6, 1773, following a vote of 15–6 in the Academy's assembly.[59] The Buffon faction did not fall silent, however. Outraged, they conspired to prevent Condorcet, who still subsisted only on a small family pension, from drawing a salary from his new office. Their opposition to Condorcet continued until 1776, by which time Condorcet had gained such overwhelming support for his tireless efforts to promote the Academy's interests that further opposition was fruitless.

The continued controversy in the Academy only served to encourage the Voltairean faction's sense of ideological struggle against the forces of darkness (absurd though it is to place Buffon in such a category), as a letter from Voltaire to Condorcet in May 1773 indicates:

> One only needs two men such as you and M. d'Alembert to conserve the source of the sacred flame from those hypocrites who want to extinguish it. And thank God! You have in Paris a great number of honest men who second you. Thus, Monsieur, never be discouraged. Once reason has put a foot in a country, one can persecute it, one can silence it for a while, but one can never chase it away. You will always be a leader of the wise. That's the best place in the world, in my view.[60]

Such rhetoric, as Voltaire no doubt intended, discouraged complacency and compromise and inspired action. Condorcet, for his part, was disposed not only to listen to Voltaire's rhetoric but also to perpetuate it and to act upon it. One of his first official duties as co-secretary, for example, was to prepare a eulogy for the famous chemist and explorer, Charles-Marie de la Condamine. A long and detailed *éloge historique*, the eulogy appeared in early 1774 and expressed clearly Condorcet's more overtly political interpretation of Fontenelle's moral idea of the sciences. In Condorcet's hands, La Condamine became the archetype for the perfect savant, guided by reason and science and compelled by them to act for the greater good of humanity. According to Condorcet, La Condamine's philosophical love of the sciences led him to direct his remarkable energies toward the most proper and virtuous ends. Quitting the military, the abode of the violent and backward nobility, La Condamine devoted himself to the sciences, "flattering himself, with reason, with the hope of being more useful to his country in this new career, and sure at least that his success would cost only the tears of humanity's enemies." The quest for knowledge and truth, not conquest, took him on voyages of discovery to the Levant, the Americas, Italy, and England, where "he was equally ready to sacrifice for the success of the enterprise his fortune, health, and life." Inspired (not surprisingly, like Condorcet himself) by a greater goal, La Condamine's "first passion was the most noble of all the sentiments: the desire to merit an illustrious name by services rendered to humanity."[61]

Condorcet's panegyric to La Condamine expressed what must have been genuinely held philosophical convictions. Voltaire, writing in February, noted that La Condamine's history was very curious and added with typical sarcasm, "They say he died in a manner very un-philosophical by putting himself in the hands of a charlatan who killed him. I well know that most men die in the hands of charlatans, be they experimentalists or otherwise."[62] Condorcet, with apparently less appreciation for sarcasm than Voltaire, insisted that La Condamine, who had died during experimental kidney stone surgery, had "died a hero of an operation to which he had submitted because of his zeal for humanity."[63]

Not content merely with rhetorical exercises expounding the virtue of public duty and service to humanity, however, Condorcet sought means to engage in public action as both intellectual and civil servant. In 1774, he anonymously published a pamphlet titled *Lettre d'un théologien* viciously attacking the reactionary abbé Antoine Sabatier's anti-*Encyclopédie*, the *Dictionnaire des trois siècles*.[64] Upon the elevation of his friend, Turgot, to the ministry in 1774, Condorcet leapt at the opportunity to engage in direct reform on his own. "Nothing more fortunate could happen to France or to human reason," he exclaimed to Voltaire two days after the momentous event. "Never has there entered into any king's council a man in whom was united such virtue, courage, disinterest, love of the public good, enlightenment, and the zeal to expand it. Since this event I sleep and awake as tranquilly as if I were under the protection of all the laws of England."[65]

Soon appointed inspector of the mint and inspector-general of navigation by Turgot, Condorcet proposed a truly prodigious program of progressive reforms. "There are things to do for the public good!" he insisted to Turgot, recommending projects to open the interior waterways of France to encourage trade, redesign the river barges, expand the canal system, standardize the system of weights and measures, initiate scientific voyages, desalinate ocean water, and reform corrupt legal procedures—particularly to suppress the use of judicial torture, proscribe religious fanaticism, and, above all, bring the murderers of La Barre to justice.[66] Turgot, exasperated by Condorcet's endless suggestions, wrote in August, "I am not responding at all, Monsieur, to all your craziness. On many points you are preaching to the converted, on others you are not capable of judging what is possible under the circumstances, and above all you are too impatient."[67]

Condorcet's zeal for action and his eagerness to defend Turgot's reforms (particularly during the Flour War) did not save Turgot's ministry, but his activities won him considerable support in the Academy of Sciences. His proposals to open the interior waterways of France, to redesign the river barges, and to expand the canal system, not to mention the major project to standardize the system of weights and measures according to a geophysical constant, necessitated increased funding for scientific and technological research—all of which would be funneled through the Academy.[68] Condorcet also prepared a report to Maurepas proposing increased authority for the Academy to coordinate scientific research throughout the kingdom. The rejection of this plan after the fall of Turgot, when so many projects were overturned, did not diminish the esteem Condorcet had earned in the Academy for his efforts. When Fouchy retired in July 1776, Condorcet renounced his claim to the office and submitted to an open election. On August 7 the academicians unanimously elected him perpetual secretary.[69]

In the context of an ideology articulated in such Manichean terms—as an apocalyptic struggle between light and darkness—Condorcet was hardly prepared to interpret the parlements' dogged resistance to Turgot's reform plans as anything other than a fundamental contradiction of his own values and beliefs. Such rhetoric expressed an uncompromising dualism and allowed Condorcet, like Voltaire, to paint a simplified picture of his enemies that dominated the historiography at least until the 1960s: the parlements were the bastions of the backward, petty, and self-interested aristocratic powers, including the Church. The monarchy's failure to support Turgot, and his tragic fall from power, convinced Condorcet to abandon the long-held Neostoic assumption that rational, scientific reforms of the sociopolitical order could best be accomplished under a monarchy. In Condorcet's view, the fall of Turgot was quite easy to explain: "All those aristocratic powers that in a monarchy serve but to weary the people and embarrass the government foresaw that their destruction or their reform would follow from a just and firm administration. . . . The courtiers sensed too well that there was nothing for them to hope in M. Turgot. . . . They foresaw the destruction of their offices, of those positions that, useless to the public order yet paid for by the people, are true vexations. . . . The financiers knew that, under an enlightened ministry, working to simplify and reform the collection of taxes, the source of their excessive opulence would soon be taken away."[70]

After Turgot's fall from power, Condorcet's writings between 1776 and 1789 subsequently became even more dualistic, and we see an unmistakable ideological shift in favor of revolutionary change. As I recount in the following chapter, his academic eulogies became more overtly political, and he attributed republican beliefs and goals to his heroes, Turgot and Voltaire, presumably to justify his own emerging republican sentiments.

Clearly, Condorcet came to believe that a republic, perhaps along the new American model, was the best alternative to monarchy, but the question of how to establish such a republic weighed heavily on Condorcet's mind well into the 1780s. In his *Vie de Turgot* (1786) he insisted that Turgot had for years believed a republic the best of all possible constitutions, and in the first addition to Voltaire's *Oeuvres* (1789) he asserted that "only a slave could prefer a kingdom to a well constituted republic where men would truly be free." Neither Turgot nor Voltaire, however, had ever seriously proposed a republican revolution: in reality "one can only choose between monarchy, aristocracy, and anarchy, and in this case a wise man can certainly prefer a monarchy."[71] Nevertheless, in the 1780s Condorcet began to suggest that an absolutist monarchy was only a stage toward the establishment of a modern republican constitution—a stage that would be passed when general enlightenment had progressed sufficiently that the people could be trusted to govern itself wisely. The duty of savants was therefore to direct public enlightenment and prepare the way for the final reformation (or revolution) of society.

NOTES

1. By ideology I mean here not a cultural structure or mentality, nor an interpretive disposition, but a clearly (though not necessarily logically coherent) articulated idea or set of ideas expressing an ideal to which the world ought to conform, or ought to be made to conform. *Philosophie* is therefore an ideology expressing a desire to make the world conform to an ideal of scientific rationality as these philosophes understood it.
2. See Gay, *Voltaire's Politics*, 331–33. There were, of course, several attempts, most notably under Calonne in 1786 (advised by Turgot's associate Du Pont de Nemours), to reinstitute most of Turgot's reforms.
3. Condorcet publicly declared his republican sentiments in a speech at the Cirque du Palais-Royal, July 12, 1791, published as "De la République, ou un roi est-il nécessaire à la conservation de la liberté?" *Oeuvres de Condorcet*, 12: 227–37.
4. See also Adkins, "When Ideas Matter."
5. See chapter 3.
6. Jonathan Israel differentiates a moderate, mainstream Enlightenment, often supported by governments, that sought to "preserve and safeguard what were judged essential elements of the older structures," with a radical Enlightenment that sought to sweep away existing structures entirely. According to this view, Voltaire's ideology of rational reform was part of the mainstream Enlightenment, but the line between moderate and radical Enlightenment ideals seems thin indeed. See Israel, *Radical Enlightenment*, 11–12.
7. Voltaire, "La voix du sage et du peuple," *Oeuvres Complètes de Voltaire*, 23: 466–71. Translation from Gay, *Voltaire's Politics*, 136–37.
8. See d'Alembert's to Voltaire, January 11, 20, and 28, and February 8, 15, and 25, *Oeuvres de d'Alembert*, 1: 51–60. To Voltaire's urging, d'Alembert responded with exasperation: "Oui, sans doubte, mon cher maître, l'*Encyclopédie* est devenue un ouvrage nécessaire, et se perfectionne à meseure qu'elle avance; mais il est devenu impossible de l'achever dans le maudit pays où nous sommes."

9. Voltaire to d'Alembert, November 17, 1760, *Oeuvres Complètes de Voltaire*, 41: 64–65. "Je n'ai guère vu d'homme plus aimable ni plus instruit; et, ce qui et assez rare chez nos métaphysiciens, il a le goût le plus fin et le plus sûr. Si vous avez plusieurs sages de cette espèce dans votre secte, je tremble pour l'*infâme.*"

10. See chapter 3.

11. Voltaire to Turgot, September 2, 1761, *Correspondance de Voltaire*, 6: 557. "Je crois même qu'il n'y a qu'un intendant qui puisse être utile. Ne peut-il pas faire réparer les chemins de traverse qui sont absolument abandonnés dans le royaume? Ne peut-il pas faire défricher des terres, dessécher des marais, encourages des manufacturers." "Les parlements ne peuvent faire que rendre justice, et la justice est souvent un grand mal; ils peuvent aussi faire des remonstrances, et ce n'est pas toujours un bien; car elles servent à décourager la nation."

12. Darrin M. McMahon, *Enemies of the Enlightenment: The French Counter-Enlightenment and the Making of Modernity.*

13. Voltaire to La Harpe, June 10, 1776, *Oeuvres Complète de Voltaire*, 50: 32. "Je vous prierai surtout de me dire où est l'intrepide philosophe M. de Condorcet. Est-il à Paris? N'est-il occupé à consoler M. d'Alembert? Ni eux ni moi ne nous consolerons jamais d'avoir vu naître et périr l'âge d'or que M. Turgot nous préparait." And to Dupont, June 15, 1, ibid., 43: "M. Turgot, le père du peuple m'encourageait . . . la destitution de ce grand homme m'ecrasse, et je vais mourir en le regrettant."

14. See chapter 3.

15. On structure as I intend it here, see Sewel, *Logics of History*, especially 124–51. See also Pierre Bourdieu, *Outline of a Theory of Practice.*

16. D'Alembert to Frederick II, September 12, 1774, *Oeuvres de d'Alembert*, 5: 352–53.

17. Arthur Wilson, *Diderot*, 597–98, 677–81.

18. The exile of the parlements in 1771 demonstrated to many the illusion of the parlements' claim to be a representative body between the people and government, leading to discussion about the old Estates-General, a body assumed to be more representative and thus more authoritative. See William Doyle, *The Oxford History of the French Revolution*, 58–59.

19. I have always been struck by the perspicacity of Peter Gay's argument that the philosophes defined themselves through work. Condorcet, for one, never seems to have rested at all. Gay, *Party of Humanity*, 116.

20. See R. R. Palmer, *The Improvement of Humanity: Education and the French Revolution*, 129–39.

21. Voltaire to Condorcet, July 11, 1776, *Oeuvres de Condorcet*, 1: 124. "Pour vous, je vous crois enfoncé dans la géométrie; je vous le pardonne, si vous faites dans les mathématiques des découvertes nouvelles, comme ont fait sir Isaac et le capitaine Halley."

22. See, for example, Frank Manuel's classic analysis of Condorcet in *The Prophets of Paris*.

23. See, for example, Voltaire to Condorcet, April 9, 1777, *Oeuvres de Condorcet*, 1: 149. "La nature m'a joué un mauvais tour, mon très-respectable philosophe. Je combats un peu contre elle, mais je serais bientôt vancu. Je veux que vous me promettiez pour ma consolation de daigner prendre ma place à l'académie des paroles, quoique vous soyez le soutien de l'académie des choses, et d'être reçu par M. D'Alembert. J'irai me présenter là-haut ou là-bas, ou nulle part avec plus de confiance."

24. Badinter and Badinter, *Condorcet*, 22–29.

25. D'Alembert to Lagrange, April 10, 1769, *Oeuvres de Lagrange*, 14: 130. Cited in Badinter and Badinter, *Condorcet*, 36. "Vous avez peut-être appris par les gazettes que nous avons reçu M. de Condorcet, la famille ayant jugé à propos de ne plus mettre d'obstacle à ce qu'il fût de l'Académie, car beaucoup de nos gentilhommes croient que le titre et le métier de savant dérogent à la noblesse."

26. See Condorcet's notes on education and religion, Bibliothèque de l'Institut, MS 884, folio 270, and Badinter and Badinter, *Condorcet*, 18–22.

27. One of the earliest letters I have seen explicitly attacking the Jesuits was to Voltaire, March 6, 1774, *Oeuvres de Condorcet*, 1, 29: "Savez-vous qu'il a été sérieusement question de rétablir les jésuites, c'est-à-dire de former une congrégation de prêtres chargés spécialement d'élever la jeunesse, et dont les ex-jésuites formeraient la plus grande partie? Nos ministres ont

eu la bonté de ne pas rire au nez de ceux qui ont proposé ce beau projet, mais après un examen sérieux ils l'ont rejeté. On s'est rabattu à former une congrégation d'éducation, dont les jésuites seraient exclus; mais ce seront toujours des fanatiques et des moines. C'est comme si les Caraïbes changeaient l'habitude d'aplatir en large la tête de leurs enfants, en celle de l'aplatir en long; ils n'en resteraient pas moins imbéciles. . . . L'agonie de la supersition sera longue, et elle a quelquefois des intervalles de vigueur qui font frémir!"

28. Admittedly, there was more to the case than popularized in Voltaire's two accounts of it. See Jean Cruppi, "Linguet et le Proces du Chevalier de la Barre," *Revue des Deux Mondes*, LXV 128 (Paris, 1895): 167.

29. D'Alembert to Voltaire, February 22, 1770, *Oeuvres Complète de Voltaire*, correspondance 14: 571.

30. Voltaire to d'Alembert, July 23, 1766, *Oeuvres Complète de Voltaire*, correspondance 12: 356–57. See also Turgot's use of the phrase *boeuf-tigre* in chapter 3.

31. There is some evidence that the plan in fact originated with Condorcet. See Voltaire to Condorcet, November 23, 1774, *Oeuvres de Complètes de Voltaire* correspondance, 17: 129.

32. D'Alembert to Voltaire, February 22, 1770, *Oeuvres de Complètes de Voltaire*, correspondance, 14: 571.

33. Condorcet to Voltaire, March 6, 1774, *Oeuvres de Condorcet*, 1: 29. "L'agonie de la supersition sera longue, et elle a quelquefois des intervalles de vigueur qui font frémir!"

34. "Biographie de Jean-Antoine-Nicolas Caritat de Condorcet par M. Arago (lue à la séance publique de l'Académie des sciences le 28 décembre 1841)," *Oeuvres de Condorcet*, 1: i–clxxi. Arago's remark that "Voltaire devint une sorte de Dalaï-Lama du monde intellectuel" appears on p. xxxvi.

35. "Discours prononcé par M. de Condorcet à l'ouverture de la séance de l'Académie Royale des Sciences, du 4 Septembre 1784." *Oeuvres de Condorcet*, 1: 448.

36. Condorcet to Turgot, January 20, 1771, *Oeuvres de Condorcet*, 1: 181. "Vous trouverez cependant dans le *Mercure* que *le moyen le plus sûr de rendre les hommes meilleurs et plus heureux est de les éclairer, et que l'estime de ceux qui éclairent les nations peut seule consoler des peines du gouvernement*. Voilà ce qu'écrivent des princes; mais ces princes, élevés par des philosophes, n'ont aucune idée de la politique noble, éclairée, honnête, de nos grands hommes d'État. . . . Je voudrais vous revoir et causer avec vous sur les malheurs de la raison et de l'humanité, sur nos espérances, s'il en reste encore, et sur les consolations que nous devons attendre du temps, qui met les hommes et les choses à leur places."

37. Condorcet to Voltaire, April 10, 1772, *Oeuvres de Condorcet*, 1: 6. "Notre ami est secrétaire perpétuel de l'Académie française. Les ennemis de la philosophie ont fait une belle défence; mais les soldats de Gédéon vaincront toujours les Madianites en les éblouissant à force de lumière."

38. Baker recounts the details of Condorcet's bid for the secretariat in "Les débuts de Condorcet au secrétariat de l'Académie royale des sciences (1773–1776)," *Revue d'histoire des sciences* 20 (1967): 229–80. See also Badinter and Badinter, *Condorcet*, 129–35.

39. MS 7464, ff. 48–52, Bibliothèque de l'Arsenal, 48v: "Il faut qu'il ait le talent d'ecrire, sinon comme l'illustre Fontenelle, du moins avec pureté, clarté, et elegance. . . . A l'egard des Eloges, il doit avoir sans doute la vérité devant les yeux, mais la vérité honnorable pour l'académiciens dont il consacre la mémoire: il doit penser qu'il n'est point juge, il n'est qu'historien: il n'employera cependant pas le même ton en parlant des vérités decouvertes, et des sistêmes hazardés; mais si'il ose proposer quelques doutes sur ces derniers, il se gardera de décides: il est sensé rapporter l'opinion de l'académie qui ne se decide que pour la vérité et contre l'erreur, mais qui permet tous les sistêmes sans en admettre aucun. Nous n'entrerons point dans les qualités du stile propre aux éloges, nous dirons seulement qu'il doit être intéressant, afin que le public entende et lise avec plaisir ces eloges, qu'il apprenne ce qu'ont faire les grands hommes qui sont morts en travaillant pour lui, et ce qui ont merité de la patrie."

40. Condorcet to Turgot, *Oeuvres de Condorcet*, 1: 197. "Vous êtes bien heureux d'avoir la passion du bien public et de pouvoir la satisfaire; c'est une grande consolation et d'un ordre supérieur à celle de l'étude."

41. Turgot to Condorcet, June 21, 1772, *Oeuvres de Condorcet*, 1: 202–3. "Quoi que vous en disiez, je crois la satisfaction résultante de l'étude supérieure à celle de toute autre satisfac-

tion. Je suis très-convaincu qu'on peut être, par elle, mille fois plus utile aux hommes que dans toutes nos places subalternes, où l'on se tourmente, et souvent sans réussir, pour faire quelque petits biens, tandis qu'on est l'instrument forcé de très-grands maux. Tous ces petits biens sont passagers, et la lumière qu'un homme de lettres peut répandre doit, tôt ou tard, détruire tous les maux artificiels de l'espèce humaine et la faire jouir de tous les biens que la nature lui offre."

42. Turgot to Condorcet, July 14, 1772, *Oeuvres de Condorcet*, 1: 207–8. "Mademoiselle de l'Espinasse m'apprend que vous travaillez sur quelque objet de littérature; je ne crois pas qu'il y ait d'indiscrétion à vous demander sur quoi . . . donnez-moi quelquefois de vos nouvelles, et de celles des choses qui vous intéressent, sans user de représailles avec moi."

43. Condorcet to Turgot, October 1, 1772, *Oeuvres de Condorcet*, 1: 208. "Je profiterai avec grand plaisir de vos remarques sur l'éloge de Fontaine, Monsieur; j'y ajouterai sûrement un mot sur l'obscurité nécessaire du compte que je rends de ses principaux ouvrages, et en même temps j'en ferai sentir l'utilité éloignée. Si vous voulez, je vous enverrai cette addition à part; vous la ferez coudre à votre copie."

44. Eulogy of Fontaine, *Oeuvres de Condorcet*, 2: 139–56. "*J'ai cru un moment qu'il valait mieux que moi*, disait-il un jour d'un jeune géomètre; *j'en étais jaloux, mais il m'a rassuré depuis*. Ce jeune homme est l'auteur de cet Eloge."

45. *Oeuvres de Condorcet*, 2: vii. "Je préfère leur histoire, plutôt que leur éloge; car one ne doit aux morts que ce qui est utile aux vivants: la vérité et la justice."

46. *Oeuvres de Condorcet*, 2: 139–56. Quotations, respectively: "Tel était cet homme, aussi original peut-être par son caractère que par son génie, que la société n'avait ni corrompu ni poli; qui porta tous les goûts de la nature et qui les suivit sans constrainte, et sans jamais connaître les passions violentes, qui ne sont l'ouvrage que de la société." "Heureux par l'étude, ayant dans la géométrie et dans la culture de sa terre, un remède sûr contre l'ennui, M Fontaine n'avait besoin ni des services, ni, malheureusement pour lui, de l'amitié de personne. Il ne cherchait point à plaire, et dédaignait de se faire craindre." "Il avait toujours regardé la douleur et la mort comme une suite nécessaire des lois générales de la nature, dont il serait absurde de se plaindre."

47. Condorcet to Turgot, October 1, 1772, *Oeuvres de Condorcet*, 1: 208–9.

48. They were Chambre, Roberval, Frénicle, l'abbé Picard, Mariotte, Duclos, Blondel, Perrault, Huygens, Charas, and Røemer.

49. *Oeuvres de Condorcet*, 2: i–viii. "Par le règlement de 1699, le secétaire doit faire les éloges des académiciens morts. M. de Fontenelle les a fait jusqu'en 1740. Ce recueil est un des livres qu'on relit le plus; et on ne le relit jamais sans y découvrir de nouveaux charmes, et sans admirer ce talent si rare d'être clair dans les choses les plus difficiles; de dire les plus communes avec finesse, et les plus fines avec cette simplicité qui les rend plus piquantes. Les savants, mort avant 1699, n'ont point eu d'éloges, et cet ouvrage manquait à l'histoire des sciences. J'ai essayé d'y suppléer."

50. *Oeuvres de Condorcet*, 2: 11. "Cette vanité malentendue nuisit à la réputation et au repos de Roberval. S'il avait étudié la géométrie de Descartes, au lieu de la combattre, il aurait été le premier parmi ses disciples. . . . Pourquoi faut-il que tant d'hommes, faits pour contribuer aux progrès des sciences, perdent, à combattre un homme supérieur, les forces qu'ils ne devraient employer qu contre la résistance de la nature? Pourquoi faut-il qu'ils sacrifient si misérablement, à une petitesse d'amour-propre, l'intérêt des sciences et celui de leur véritable gloire? . . . Dans ses méthodes, presque toujours obscures et embarrassées, on ne peut cependant méconnaître l'empreinte du génie."

51. *Oeuvres de Condorcet*, 2: 23. "Moins sensible à l'intérêt de sa réputation qu'au plaisir de donner à sa patrie un homme qui lui serait utile, il ne fut point frappé de la crainte d'y avoir un rival occupé du même object, et qui pouvait être dangereux pour sa gloire."

52. *Oeuvres de Condorcet*, 2: 81–82. "Heureusement pour les habitants de Copenhague, Frédéric IV était supérieur à ce préjuge, si commun dans les cours, que les savants sont incapables des places d'administration; comme si l'habitude de chercher la vérité ne pouvait pas tenir lieu de la routine qui s'acquiert dans les emplois subalterns. . . . Si pourtant l'on prend l'esprit d'intrigue pour celui des affaires, et l'art de tromper ou d'opprimer les hommes pour celui de les gouverner, on a raison de croire que les savants n'y sont pas propres, et qu'une âme qui s'est longtemps nourrie de l'amour de la vérité et de la gloire, ne peut guère ni sentir la

nécessité, ni prendre l'habitude de ce mélange de fausseté et de bassesse qu'on décore du nom d'habileté."

53. *Oeuvres de Condorcet*, 2: 53. "Ce sont eux [Claude and his brothers] principalement qui ont inspiré à Colbert de protéger les sciences d'une manière si flatteuse pour les savants, et si honorable pour lui-même. Le cardinal de Richelieu n'avait vu, dans les gens de lettres, que des hommes utiles à sa gloire: Colbert, dirigé par Perrault, les traita comme des hommes utiles à l'humanité, dont la gloire honore leur pays et leur siècle. . . . Cette justice, rendue aux talents par la puissance, a rappelé les gens de lettres à la dignité de leur état, qu'ils avaient oubliée."

54. See Voltaire to Condorcet, December 5, 1772, December 24, 1772, March 1, 1773, *Oeuvres de Condorcet*, 1: 21–24.

55. Lagrange to Condorcet, October 19, 1773. Cited in Badinter and Badinter, *Condorcet*, 108.

56. Badinter and Badinter, *Condorcet*, 108.

57. Cited in Badinter and Badinter, *Condorcet*, 108.

58. Bibliothèque de l'Arsenal, MS 7464, 52r: "Sans doute l'intérêt personnel, ou même, sans supposer ce motif, une certaine inquiétude naturelle à certains esprits, leur fait toujours voir quel qu'avantage illusoire dans le changement; mais depuis cent ans l'académie des sciences s'est bien trouvée de n'avoir qu'un sécretaire; et quand après de 80 ans l'illustre Fontenelle déposa le fardeau de ses fonctions: cet homme qui l'histoire des sciences a seule immortalisé, cet homme qui peut être en valoir plusieurs autres par l'universalité de ses connoissances et par un coup d'oeil vrayment philosophique, fut cependant remplacé par un seul homme."

59. On the details of the controversy over Condorcet's election, see Baker, "Les débuts de Condorcet."

60. Voltaire to Condorcet, May 23, 1773, *Oeuvres de Condorcet*, 1: 16–17. "Vous êtes un vrai philosophe, Monsieur, c'est-à-dire un vrai sage, et vous rendez la philosophie bien aimable par les grâces de votre esprit. Il ne faut que deux hommes comme vous et M. d'Alembert pour conserver le dépôt du feu sacré que tant d'hypocrites veulent éteindre; et Dieu merci! Vous avez dans Paris un très-grand nombre d'honnêtes gens qui vous secondent. Ainsi, Monsieur, ne vous découragez jamais. Quand la raison a mis une fois le pied dans un pays, on peut la persécuter, on peut la faire taire pour quelque temps; mais on ne peut la chasser. Vous serez toujours à la tête des sages. C'est la plus belle place du monde à mon gré."

61. *Oeuvres de Condorcet*, 2: 158–204. Quotations, respectively: "Il quitta l'état militaire pour se livrer aux sciences, se flattant, avec raison, de l'espérance d'être plus utile à son pays dans cette nouvelle carrière, et sûr du moins que ses succès ne coûteraient de larmes qu'aux ennemis de l'humanité et des talents." "Pour vaincre les difficultés que de telles circonstances devaient faire naître à chaque pas, il fallait un dont l'activité crût avec les obstacles, qui fût également prêt à sacfrifier au succès de son enterprise sa fortune, sa santé, et sa vie." "Mais il eut le courage d'y renoncer, pour aller passer dix ans dans les déserts du Pérou; ce qui prouve du moins que sa premier passion était le plus noble de tous les sentiments, le désir de mériter un nom ilustre par des services rendus à l'humanité."

62. Voltaire to Condorcet, February 25, 1774, *Oeuvres de Condorcet*, 1: 25. "On dit qu'il est mort d'une manière très-anti-philosophique, en se mettant entre les mains d'un charlatan qui l'a tué. Je sais bien que la plupart des hommes meurent entre les mains des charlatans, soit empiriques, soit autres."

63. Condorcet to Voltaire, March 6, 1774, *Oeuvres de Condorcet*, 1: 26–29. "La Condamine est mort en héros d'une opération à laquelle il s'est soumis par zèle pour l'humanité."

64. Although denying authorship, Condorcet admitted, tongue-in-cheek, to Turgot that he found it very agreeable and liked "the spirit in which it has been composed." See Condorcet to Turgot, July 1774, *Oeuvres de Condorcet*, 1: 242–4. The pamphlet was so caustic and anti-religious Voltaire feared to be implicated as the author and vigorously denied having anything to do with it. See Voltaire to Condorcet, August 20, 1774, ibid., 1: 40–42. "L'abbé qui a fourni le canevas de cette brochure n'a fourni que des pierres avec lesquelles on lapidera les philosophes. Je veux bien être lapidé pour sauver d'honnêtes gens, mais je ne veux pas mourir injustement et inutilement. Il n'est pas juste qu'un abbé Sabotier, le plus vil des scélérats, compilateur du système de Spinosa (et non de la vie de Spinosa, comme le dit la lettres du

Théologien), un soi-disant homme de lettres qui n'a fait que des vers aussi infâme que plats, un domestique qui a volé ses maîtres, un fripon échappé des cachots où il avait été renfermé à Strasbourg; il n'es pas juste, dis-je, qu'un tel homme puisse être, avec quelque apparence de raison, la cause du malheur des hommes le plus respectables et du mien. On doit me laisser mourir en repos. En un mot, je ne suis pas l'auteur de la lettre d'un Théologien."

65. Condorcet to Voltaire, July 22, 1774, *Oeuvres de Condorcet*, 1: 36–38. "Vous savez sans doute la nomination de M. Turgot. Il ne pouvait rien arriver de plus heureux à la France et à la raison humaine. Jamais il n'est entré dans aucun conseil de monarque d'homme qui réunît à ce point la vertu, le courage, le désintéressement, l'amour du bien public, les lumières et le zèle pour les répandre. Depuis cet événement, je dors et je me réveille aussi tranquillement que si j'étais sous la protection de toutes les lois de l'Angleterre."

66. Condorcet to Turgot, July 1774, *Oeuvres de Condorcet*, 1: 243. "Qu'il y a de choses à faire pour le bien public! Proscrire le fanatisme et faire justice des assassines de La Barre; assigner pour chaque crime une peine légale; supprimer la question [torture] et des supplices barbares, trop éloignés de nos moeurs; établir enfin un tribunal où le particuler insulté par un magistrat, ou qui aurait un procès avec lui, serait jugé par d'autres que par les confrères de son adversaire." See also Badinter and Badinter, *Condorcet*, 110–57.

67. Turgot to Condorcet, August 17, 1774, letter cxlii, *Correspondance de Condorcet et de Turgot*, 192.

68. See chapter 3 for a more detailed discussion of the scientific activities of Turgot's ministry.

69. Baker, "Les débuts de Condorcet," and Badinter and Badinter, *Condorcet*, 122–23, 135.

70. "Vie de Turgot" [1786], *Oeuvres de Condorcet*, 5: 110–11. "Tous ces pouvoirs aristocratiques, qui, dans une monarchie, ne servent qu'à fatiguer le peuple et à embarrasser le gouvernment, prévoyaient que leur destruction ou leur réforme serait la suite d'un système d'administration juste et ferme. . . . Les courtisans sentaient trop bien qu'ils n'avaient rien à espérer de M. Turgot. . . . Ils prévoyaient la destruction de ces charges, de ces places qui, inutiles à l'ordre public, et cependant payées par le peuple, sont de véritables vexations. . . . Les financiers savient que, sous un ministre éclairé, occupé seulement de simplifier et de réformer la perception de l'impôt, les sources de leur excessive opulence allaient bientôt tarir."

71. "Avertiseements insérés par Condorcet dans l'edition complète des Oeuvres de Voltaire," *Oeuvres de Condorcet*, 4: 493. "Il n'y a qu'un esclave qui puisse dire qu'il préfère la royauté à une république bien constituée, où les hommes seraient vraiment libres . . . mais cette république n'exite point et n'a jamais existé. On ne peut choisir qu'entre la monarchie, l'aristocratie, et l'anarchie; et, dans ce cas, un homme sage peut très-bien donner la préférence à la monarchie."

Chapter Five

The Coming of the Tenth Epoch: The Idea of the Sciences to the Revolution of 1789

> In a word, are men approaching a state where all will have the enlightenment necessary to conduct themselves according to their own reason in the normal affairs of life and maintain themselves free from prejudices, to know well their rights and exercise them according to their opinion and conscience, where all will be able, by the development of their natural faculties, to provide for their needs, where, finally, stupidity and misery will no longer be but accidents and not the normal state for a portion of society?
>
> —Condorcet, *Esquisse d'un tableau historique des progrès de l'esprit humain*, dixième epoch

Condorcet wrote the *Esquisse d'un tableau historique des progrès de l'esprit humain*, quoted above, probably in October 1793 while in hiding during the height of the Terror. In this work, published posthumously, Condorcet articulated what we now generally consider the greatest ideal of the Enlightenment: limitless progress toward infinite perfectibility through the power of individual human reason liberated from outside constraints and controls—emancipation through education.[1]

But at the time, Condorcet's work in the *Esquisse* actually marked a significant rupture in the idea of the sciences that I have hitherto traced in this book, a democratization of a previously highly elitist idea. Of course, this rupture was a long time coming, dating at least from 1776, as this chapter recounts. From the fall of Turgot and until the end of life, Condorcet worked out his reformulation of the century-old moral idea of the sciences meticulously and systematically in an effort to find an alternative to the method of reform in which he had obviously lost hope. The idea of emancipation through education—as expressed in the *Esquisse*—followed a detailed plan for establishing an enlightened democratic polity through revolutionary constitutional and educational reform that Condorcet developed in the 1790s and with which almost no one else fully agreed—and for which he despaired, struggled, and eventually died. Even when others (particularly among his colleagues in the National Assembly, as we shall see below) agreed in principle with Condorcet's prescriptions for creating an enlightened egalitarian polity, few were willing to push as hard as Condorcet or to abandon immediate political goals in order to obtain such an end. The fact that Condorcet's idealism is more often than not taken as general for his age is a credit to the influence of his ideas on the succeeding generations. The reality is that Condorcet went his own way philosophically, ideologically, and even politically on the eve of the Revolution. From the weight of his strong reputation, inexhaustible energy, and towering intellectual, not to mention physical, stature (he stood at around six feet in an age when the average height was five feet three inches), he was able to influence in some way just about everyone with whom he worked, but the result of the relentless pursuit of his own goals was his political isolation—and, finally, his proscription and death during the Terror of 1793–1794.

The fall of Turgot from power in 1776 clearly had a profound affect on Condorcet, who had eagerly served in Turgot's ministry as both advisor and civil servant. As he wrote to Voltaire soon after, Turgot's dismissal was the end of a dream that, though Condorcet did not express it that way, had begun with Sorbière in the seventeenth century.[2] Despairing of the ability of the monarchy to affect the reform of France, Condorcet sought alternative means to bring about this desired transformation. He found his alternative to an absolutist monarchy in an educated or enlightened republic, which his plan for public instruction was ultimately intended to create. This chapter explores Condorcet's post-Turgot philosophical pivot—his decision to abandon absolutism and promote the idea of a modern republicanism in an effort to make reason, or the moral idea of the sciences, operate freely of political force, independently and self-evidently, so long as proper education was distributed widely enough (and Condorcet had very particular ideas about what constituted "proper" education). As Johnson Kent Wright argues in his study of the abbé Mably and *classical* republican thought, Condorcet in this period is associated with the formation of *modern*—that is, a specifically *anti-classical*—republican thought.[3] As we shall see, Condorcet brought about the culmination of his plan to form a modern republic in 1792. Everything after made up the final act of the tragedy.

On April 20, 1792, sixteen years after the fall of Turgot, Condorcet stood as head of the Committee on Public Instruction in the Legislative Assembly to deliver a report on his committee's education program for the nation. Just a week earlier he had put the finishing touches on his master plan for public education in France, the *Cinq mémoires sur l'instruction publique*, and the report to the assembly briefly outlined the program therein and called for immediate legislation to implement it.[4] Condorcet seemed on the eve of a great success. Only by universal enlightenment (that is, universal public instruction in "positive knowledge"), he argued, could French society successfully make the transition from the failed regime to a new and stable order of popular sovereignty and representative government in which social and political tranquility were guaranteed by the proper application of reason to the social and political spheres. Condorcet's plan for a French national educational system was intended to establish firmly the foundations for an enlightened republic by creating the conditions whereby, as he noted, "a greater number of men become capable of fulfilling well the functions necessary to society, and that the progress resulting from enlightenment opens an inexhaustible source of relief for our needs, remedies against our problems, and the means of providing individual happiness and common prosperity."[5]

Condorcet's plan was the culmination of his thinking in the post-1776 era on the means of establishing a feasible alternative to absolute monarchy. The Revolution provided him with the opportunity and the environment to develop and propose his ideas on public instruction and representative government—but, in reality, these ideas were not a radical departure from his pre-1789 thought. The transformation of the moral idea of the sciences occurred incrementally. Nor were the underlying principles any different from the moral philosophy of the sciences he had adopted and elaborated in the 1770s (see chapter 4). Condorcet's plan would, in short, reify as a state project what he saw as a historic struggle of reason to overcome ignorance and officially recognize the public role of savants as the guardians of social and political order.

Scholars have noted that from the mid-1770s to the mid-1780s, Condorcet avoided active involvement in politics and government, and he concentrated his energies on intellectual endeavors. The result of his labor was the development of his "social mathematics," and especially his probability analysis of voting procedures, which he called "political arithmetic."[6] Condorcet no longer worked in a government ministry, but his scholarly work was politically engaged and had revolutionary implications long before the Revolution became a reality. The objective of Condorcet's analysis was not academic but it was intended to provide a blueprint for the transformation of society. The problem with participatory government, Condorcet realized, was with how to ensure that the public made good decisions, that what he termed the "probability of decisions" in any act of political or judicial decision-making was as close as possible to a moral certitude, thus bringing about the best or most just outcome. Since, Condorcet wrote, "the idea of justice, of right, is necessarily formed in the same manner in all sensible beings capable of the necessary operations," and "since every being reasoning correctly will be conducted to the same ideas in morality as he will in geometry," moral principles were as certain and invariable as mathematical proofs.[7] (Was Condorcet aware of how closely he echoed his philosophical forebear, Sorbière, who had once said, "I avow to you that Moral Science and Mathematics come together, and that Physics is an Ocean to which all our knowledge will return"?[8]) To paraphrase, an objectively determinable "correct" outcome of any question of justice and law could, therefore, be determined, at least to a moral certitude (that is, to a mathematically determined threshold of certainty, below which any incertitude was negligible), assuming that the decision-makers were reasoning correctly and not subject to prejudice, unfounded opinion, or ignorance. Even when the reasonable objectivity of the decision-makers could not be assured, Condorcet believed, mathematical analysis could be used to ensure that whatever decision was reached would approach as close as possible in that circumstance to a moral certainty by weighing the decision (or vote) of each decision-maker according to his or her level of education, knowledge, and expertise, which Condorcet saw as an indicator of proper reasoning capability. As the general level of enlightenment dropped, however, the probability of decisions being correct correspondingly diminished. The inverse implied, more optimistically, that the greater the number of enlightened decision-makers, the closer decisions would approach a moral certainty.

Condorcet's rather remarkable theory suggested that enlightenment ought to be spread to the greatest number of people possible, thus increasing the probability that political, moral, and legal judgments were correct and just. Successful social and political reform followed directly from progress in the arts and sciences, and from universal education in these arts and sciences—education directed by the nation's enlightened elite who themselves would be drawn from an exponentially increasing pool of educated citizens. In the Tenth Epoch of history, the coming of which Condorcet prophesied in his posthumous *Esquisse*, the arts and sciences will have reached such a high level that it will have become a moral certainty that everyone will make correct judgments on practically every issue all the time.[9]

From 1776 to 1792, Condorcet's intellectual and political work converged in an attempt to systematize, justify, and finally institute a philosophically acceptable alternative to absolute monarchy. The ultimate denouement of his plans and Condorcet's final agony during the Terror signaled the end of the *parti philosophique*'s Enlightenment, although not, of course, forever.

The reasons for Condorcet's failure during the Revolution are closely tied to the uniqueness of his ideas. For no obvious reason, after all, would Maximilien Robespierre, the rhetorician of the Terror who ascended to power with the Committee of Public Safety in July 1793, have opposed Condorcet's revolutionary ideas. In their ultimate commitment to republicanism as a political ideal, the two men were seemingly of like mind. Why, then, did Condorcet end up proscribed and forced into hiding, later to be arrested as a fugitive and found dead, probably at his own hand, in his prison cell?

Of course, many people met their untimely ends during the Terror of 1793–1794. In that way Condorcet was no different from thousands of others, Robespierre included. But Robespierre went to his death for particular reasons or decisions, as did Condorcet—his proscription was not random or haphazard, even if we might judge it unjust. The dramatic events that resulted in the fall of Condorcet—"the last of the philosophes," as historians have traditionally dubbed him—highlight a kind of culture war in the radical camp of the Revolution itself. The struggle was over the goals and meaning of the Revolution. This conflict of ideas was not initially between factions, between the Girondins and the Montagnards, or between moderates and radicals, even though popular history often represents it that way. On the one side were the ideals of the liberal Enlightenment, and on the other the ideals of national unity. No particular persons or factions had a corner on these ideals, for they were held by just about everybody in the National Assembly to some degree—they formed the intermingled, rarely systematic spectrum of political ideas that made up the mental dispositions of the Revolutionary political elite. Equality, progress, reason, and freedom were on the lips of most representatives in those days, but ultimately deeper than those ideals was the desire for national unity, for "fraternity," expressed by the language of patriotism that drove political discourse well before 1789.[10] In the tensions and threats of the early 1790s, as France was invaded from the east by foreign powers, the ideals of national unity won out over the ideals of the Enlightenment. The issue that moved national politics by 1793, and that devolved into the bitter factional struggle between the Girondins and the Jacobin-led Montagnards, was who would control the plan for establishing and directing national unity. Condorcet went down, as we shall see in detail below, in large part because he stood between those engaged in this struggle, not because he joined a faction.

The debates on national education in the National Assembly, particularly between 1791 and 1793, formed an important locus of the struggle over ideas—and Condorcet was at the center of the education debates. As anyone interested in, or perhaps outraged by, contemporary American debates on public education realizes, any system of or policy on national education represents far more than a simple desire to teach the youth of a nation its letters and numbers. Control over national education, after all, could allow a faction, a party, or the proponents of a particular ideology to affect or even transform public consciousness, and thus to place a stamp on a national identity.[11] Debates about public education are therefore ultimately about the exercise of political power. Up to his proscription in 1793, and even after his death in 1794, Condorcet and his ideas played a central, although mostly negative, role in the education debates of the French Revolution.

The standard interpretation of the fall of Condorcet is this: By 1792, Condorcet, a member of the social and intellectual elite, was drawn into the sphere of the "moderate" Girondin faction. As head of the Committee of Public Instruction in 1792, Condorcet's National Education plan constituted a thinly veiled attack on Jacobin principles.[12] As a result, the Mountain, in which the Jacobins held ideological ascendancy, buried the plan, asserting bogusly that it was elitist and undemocratic. Condorcet had also managed an appointment to the Constitutional Committee in September 1792, however, and he used this position further to attack Jacobin power. The resulting constitution of February 1793 sought to limit Jacobin power at the same time that it sought to make constitutional the basic features of his education plan. The Girondin deputies, therefore, vigorously approved it. The Mountain, however, denounced the constitution and proposed one of their own in late May. In order to forestall opposition, the Jacobins then participated in the purge and later execution of many of the Girondin leaders from the Assembly in early June. In apparent frustration, Condorcet hastily took up the pen to denounce the Jacobins in print, thus effectively signing his own death warrant. In early July the Committee of General Security ordered Condorcet's arrest. Forewarned, he went into hiding until forced to quit his refuge in late March 1794. Discovered and arrested just south of Paris on March 27, he was thrown into prison and discovered dead in his cell on the 28th—legend has it of self-administered poison.

Readily available evidence contradicts both the reality of a concrete factional struggle in the Assembly between the Montagnards and Girondins before 1793 and the idea that Condorcet had the support of the Girondins and the opposition of the Jacobin-led Mountain.[13] In fact, nearly everyone opposed Condorcet's plan when it came up for discussion in December 1792, but the majority in opposition were "moderates" as then defined, and several of the most strident of those who spoke against the plan were Girondins.[14] Although signs of factional division were beginning to appear in December 1792 on the side of the Mountain, none spoke out against the plan in the Convention—true, Robespierre dismissed the plan in his journal, the *Lettres à ses commettants*, but he said nothing during the December debates.[15] As it turned out, the only defender of the plan to speak in the Convention was Gilbert Romme, a member of Condorcet's Committee of Public Instruction but also a member of the Mountain.

As R. R. Palmer showed in his still essential study of education in the French Revolution, the substance of the opposition to Condorcet's plan was that it was "elitist" and failed to provide for "moral and civic training." By the end of the December debates, the Convention could come to no agreement about education and postponed the issue.[16] All succeeding plans through 1795 sought to retain some elements of Condorcet's liberal education philosophy while incorporating the patriotic education judged lacking in Condorcet's plan. None was implemented successfully.

Although sympathetic to Condorcet, Palmer ultimately accepts the charge of elitism leveled against Condorcet's education plan. In Palmer's view, Condorcet himself may have professed a faith in democracy and equality, but that faith was at odds with his belief in the sciences, in the moral idea of the sciences, and in the ideals of *philosophie* that reigned supreme in his education plan. In other words, Palmer perpetuates the idea that the *philosophie* of Condorcet, true to its Neostoic roots, was inherently elitist and undemocratic.[17] But is the division really between democracy and an elitist *philosophie*? In a word, no. Condorcet's education plan—indeed, his whole political philosophy—became liberal and democratic. It was also allergic to national identity formation, civic indoctrination, and "republican cathechisms." And that is where its trouble began.

Most French education theorists throughout the second half of the eighteenth century generally agreed that education should be nationally directed with the goal of fostering patriotic sentiment and thus of building a national identity and common notions of citizenship.[18] As historians have shown, French moralists and commentators, particularly after the demoralizing defeats of the Seven Years' War, were obsessed with the regeneration of the *patrie* to reestablish its prestige among the European states.[19] During the Revolution, the initiators of the various national education reform plans continued to pursue the goal of national identity formation—except for Condorcet. In fact, the first result of the education debates during the Revolution was the report that Charles-Maurice de Talleyrand presented to the Constituent Assembly in 1791.[20] Talleyrand, like many education reformers of his day, insisted that, in order to foster a common national identity, children should be taught uncritical and patriotic reverence of the Constitution. The very nature of the French Constitution was under dispute in 1791, however, and so long as the deputies could not agree on such a fundamental issue, they could not agree on a national education plan. Talleyrand's plan was shelved.

Condorcet, heading up the Committee on Public Instruction, put the most detailed and intellectually consistent national education plan of any proposed during the Revolution before the Legislative Assembly on April 20, 1792.[21] The French declaration of war against the Austro-Hungarian Empire that very day, however, gave Condorcet's opponents a pretense to postpone further discussion of the plan until they had studied it more thoroughly. When the plan was proposed again in December in the newly formed Convention, the deputies, as noted above, rejected it for its apparent "elitism"—which is to say, for neglecting patriotic education and seeking to establish a politically independent body in charge of curricula. Soon after, more explicit factional divisions began to form in the Convention. The Jacobins, therefore, rejected the Sieyès-Daunou-Lakanal plan of June 1793, which was highly derivative of the Condorcet plan, because many Girondins supported it. In retaliation, Girondin representatives in the Convention refused to adopt the Jacobin-supported le Peletier plan of July 1793.

Despite the failure of the Revolutionary government to adopt Condorcet's plan, Condorcet's ideas were at the heart of education reform proposals throughout the debates in the Constituent and Legislative Assemblies, as well as in the Convention. Talleyrand had sought Condorcet's expertise, and the abbé Sieyès collaborated with Condorcet on the *Journal d'instruction sociale* (1793), which sought to build a consensus regarding education reform. Even the le Peletier plan rehashed Condorcet's ideas in a form more acceptable to the then-ascendant Jacobins.[22] So why force Condorcet's plan to a crushing defeat?

Perhaps the least influenced by immediate political exigencies or petty ambitions, Condorcet's plan emerged consistently from his philosophy of enlightenment, as it had developed over the course of the 1770s and 1780s when he had transformed the moral idea of the sciences to suit a postabsolutist state. This does not mean Condorcet lacked ideological and even political goals that he hoped to achieve with his education plan. Through his plan he sought to prevent the influence of any particular faction. The plan endorsed no particular national identity, and it rejected the notion of fostering patriotic sentiments acceptable to any faction. Through his refusal of patriotic education and his attempts to undermine factional ideology, Condorcet promoted the development of a free and democratic culture that recognized no ultimate authority other than reason itself—which Condorcet considered freely available to anyone instructed to think critically. Education, or enlightenment, Condorcet professed, could provide a point of difference from power, and thus prevent tyranny in its many forms by granting agency to each citizen as an independent thinker.

Condorcet's rejection of what we could term *patriotic education* set him apart from most French education theorists in the eighteenth century, from Montesquieu and Duclos to his own mentor, d'Alembert.[23] Condorcet asserted that he was far closer in thinking to Turgot, as he wrote in his biography of Turgot in 1786.[24] Turgot, during his ill-fated tenure as Comptroller-General of Finances from 1774 to 1776, made universal education part of his strategy for transforming the ailing monarchy into what he termed a "legal despotism," in which an enlightened public and an absolute monarch worked together to establish a rational social order.[25] Despairing of reforming the monarchy after 1776, however, Condorcet transformed the elitist Neostoic elements of *philosophie* in order to create a blueprint for democracy. To understand Condorcet's intellectual transformation, it is necessary to go back to 1776.

The fall of Turgot in 1776 necessitated Condorcet's resignation from administrative office as inspector of the mint. This did not signal a withdrawal from public activism, however, but merely a change in tactics: Condorcet became an ever more vocal political critic as well as an antislavery activist and pamphleteer. Under Voltaire's influence, he had come to articulate, and thus also to perceive, the philosophic cause in the language of struggle, as a historic battle between the forces of good and evil. Such an ideology did not lend itself to retreat or to compromise with the enemy. As a result, his despair at the fall of Turgot produced an even fiercer and more daring attitude.[26] When Condorcet tendered his resignation to Maurepas in October, it was with deep regret and anger over the loss of an opportunity to act for what he believed was a greater good.[27] Although Maurepas compassionately decided to allow Condorcet to remain in his comfortable apartments at the Hôtel de la monnaie in order to spare his ill and aging mother, whom Condorcet had brought to live with him in 1775 (a difficult move), Condorcet held a spiteful grudge against the minister who he believed had betrayed Turgot and the philosophic cause. Condorcet incited Maurepas's anger in 1777 first by adamantly refusing to compose a eulogy for Maurepas's brother-in-law, an academician *honoraire* (and thus, in Condorcet's estimation, not a true savant), and second by writing what amounted to a combined critique of Maurepas and defense of Turgot—thinly disguised as a eulogy of the sixteenth-century *politique* leader Michel de l'Hôpital (Condorcet submitted the eulogy to the *Académie Française*'s annual prize competition).

D'Alembert sought to convince his impetuous friend to reconsider making Maurepas a personal enemy.[28] Voltaire, who increasingly saw Condorcet as his intellectual successor, eagerly sought Condorcet's election to the *Académie Française*. But Maurepas could easily prevent Condorcet's entrance into the Academy, especially since the *dévots* in the Academy already detested Condorcet for his anti-religious writings.[29] Voltaire's dying wish that Condorcet take his place as one of the forty immortals went unanswered, however, as Condorcet, in apparent bitterness, was unwilling to change his hard rhetoric.[30] Not only did he lose the prize competition to a now little-known writer, the abbé Rémy, Condorcet also found himself barred from the Academy so long as Maurepas lived.[31]

Although the eulogy of l'Hôpital was Condorcet's longest and most polished eulogy to date, it could hardly have been mistaken for a politically innocent or uncontroversial piece.[32] In Condorcet's hands, the famous *politique*—and, indeed, philosophically Neostoic—l'Hôpital clearly represented Turgot. As a man who refused to compromise his enlightened principles for political expediency, l'Hôpital also served as a mirror-reverse of Maurepas. Similar to Turgot, l'Hôpital had served in public office "amid the most violent fanaticism [the Religious Wars]" in order to make heard "the voice of reason and humanity." In the midst of a "long war against the enemies of public peace," he had bravely stood for the philosophic cause: "Enlightened by his experience, by study, and by long reflection, could he not sense how he must hasten to destroy those abuses that voicelessly attacked the public morality and national character?" Realizing that the stability of the state depended on a strong monarchy, he vigorously supported the king—but a king who reigned only to uphold the "rights" of the nation: "Surrounded by anarchy and revolt, he defended with courage equal to the task the authority of the king and the rights of the nation." Unlike Maurepas, however, "the corruption of his century and the intrigues of the court compromised neither his integrity nor his sincerity." As the head of state finances, l'Hôpital, like Turgot, believed that "the funds raised by state taxes cannot legitimately be employed but for the advantage of the people who have paid them." He therefore rejected all use of taxes that did not have as their object the public good, thus earning the ire of many vested interests. "But we do not complain at all of having earned the hatred of the people's enemies," wrote Condorcet, overtly linking himself and his philosophic cause historically to the *politiques*. "To a strong soul this hate is good. It is the most obvious proof that one has served the nation." As opposed to Maurepas, l'Hôpital would not fall silent in the face of parlementary intrigue and resistance, but rather sought to correct and enlighten the magistrates about their true duties: "The chancellor must never lose sight of the fact that the magistrates have been instituted for the people. . . . If fear, baseness, avidity, and partiality corrupt the purity of their judgments . . . if the magistrates lower themselves to be the instruments of the passions of powerful men, or to be complicit in their intrigues; if they neglect their useful functions in order to aspire to a semblance of power they cannot obtain except at the expense of public prosperity, then they must find in their chief a censor more occupied to enlighten them than punish them."[33]

The use of the eulogy as a political soapbox was not a great leap for Condorcet from its more traditional use as moral rhetoric. Even in his first eulogies Condorcet was not afraid to go beyond mere panegyric in order to express his own political or philosophic opinions, and he continued this practice in the late 1770s and 1780s (see chapter 4).[34] In his eulogy of Linnaeus in January 1778, for example, Condorcet highlighted the political lessons one learned, oddly enough, from the famous botanist: according to Condorcet, Linnaeus's research had led him to the conclusion that a man of science (*homme physique*) must not try to manipulate nature as a bad king might use his subjects. "If a scientist abuses his power," wrote Condorcet, "he learns soon enough from the resistance of his research subjects that kings have been established for the people, and not the people for the kings."[35] The idea that the sciences themselves led the savant toward the realization of invariable political and moral truths would not have been foreign to Fontenelle, of course, but with Condorcet this notion took on distinctly ideological, and increasingly radical, tones.

Condorcet did not confine his forays into the public arena to eulogies. During the first mesmerism controversy of 1779–1780, he had the opportunity to defend the authority of the Royal Academy of Sciences to act as the supreme arbiter of science in France.[36] Outraged by the Academy's (and Condorcet's) judgment of Jean-Paul Marat's *Découverts sur le feu, l'électricité et la lumière* (1779) as pseudoscience, the mesmerists attacked the Academy's right to judge and decried its esprit de corps, which gave preference to scientific works following a supposed proper—that is, Newtonian—methodology.[37] Jacques-Pierre Brissot, the future political ally of Condorcet, wrote disdainfully, "The moderns [by which he meant the Newtonian academicians] have introduced into the empire of the sciences a kind of elective aristocracy." This aristocracy, Brissot continued, foreshadowing his later political opposition to Condorcet's education plan in the Revolution, gave itself the unjust right to judge "productions of genius"—an authority that ought to belong only to public opinion.[38]

To Condorcet, however, defending the authority of the Academy to judge scientific knowledge was more than esprit de corps. Although his argument that experts alone should judge scientific works and that the Academy of Sciences represented the highest body of such experts was perfectly reasonable to him on its own grounds, Condorcet's argument was not independent of his desire to assert a public leadership role for savants. To his mind, the academicians did indeed form a sort of "elective aristocracy," or rather a natural elite, but that was precisely because their authority was not derived from unfounded opinion but from the objective, expert judgment of their peers—a point he lost no time explaining:

> The academies have two incontestable uses: The first is to be a barrier always opposed to charlatanism of all sorts . . . the second is to maintain in the sciences good methods and to prevent that any branch of the sciences be absolutely abandoned. They have a very important third use, which is to make savants independent of popular opinion. A chemist, an anatomist, or a mathematician who is a member of an academy has no need to run around like a charlatan in order to enjoy from ignorant people the reputation of a savant. It is the result of his works that merit him celebrity or glory.[39]

Condorcet's philosophic cause had not at all abated, but had only intensified with the fall of Turgot, and it prompted Condorcet's defense of the intellectual and moral authority of the sciences and of savants. Condorcet's controversial election to the *Académie Française* further revealed that the attitude of ideological struggle that had come to characterize the Voltairean faction since the mid-eighteenth century continued to enliven their, or at least Condorcet's, rhetoric. With the death of Maurepas in November 1781, and the death of the academician Bernard-Joseph Saurin one month later, the path for Condorcet's final acceptance into the Academy was open. Ironically, his opponent for the seat was once again the astronomer Bailly, still supported by Buffon. By all accounts the election campaign was even more contentious than the earlier one for perpetual secretary of the Academy of Sciences, and d'Alembert expended the last of his energies in all manner of machinations, seductions, and deceit to secure the seat for his protégé. On January 21, the day of the election, there was so much ill will between Buffon and d'Alembert that the former refused even to appear at the Louvre, where Condorcet won by six votes.[40]

Clearly encouraged by his victory, Condorcet used his acceptance speech to give voice to his vision. Tradition demanded the sort of empty, self-congratulatory rhetoric that Voltaire joked about in his *Lettres philosophiques*.[41] Rather than play by the rules, however, Condorcet gave an overly idealistic and probably unwelcome harangue on the imminent victory of reason over ignorance if only the learned remembered their duty to enlighten mankind—which obviously Condorcet would not let them forget. "See, now," he commanded, "from one end of Europe to the other, enlightened men join all their efforts together for the good of humanity and turn toward this object alone all their powers with a courage and concert never before seen in any century."[42] As knowledge and science progressed, he announced, the human spirit would advance beyond all boundaries, and the darkness of ignorance and error would finally be abolished: "Humanity will no longer see those periods of alternate darkness and light to which for so long we believed nature had condemned it. It is no longer in the power of men to extinguish the flame ignited by genius, and only a planetary catastrophe could bring back the shadows."[43]

Despite what even in 1781 must have seemed clichéd rhetoric, Condorcet's speech does reveal the new direction of his thoughts about education. The last great intellectual barrier to the victory of *philosophie*, Condorcet asserted, was in the moral sciences. Despite common assumptions to the contrary, he was convinced that the principles of the social world were as susceptible to rigorous analysis as the objects of the physical sciences.[44] "They [the sciences] would be the same for a being that, as a stranger to our species, would study human society as we study that of beavers or bees. . . . Being supported like the physical sciences by the observation of facts, they [the moral sciences] ought to follow the same method, acquire a language equally exact and precise, and attain the same degree of certitude."[45] If progress in the moral sciences tended to be slower than in the physical sciences, Condorcet argued, it was because of the difficulty of overcoming entrenched social traditions and prejudices, as well as the particular interests and passions of discrete individuals, in order to arrive at a truly collective reasoning.[46] But such ignorance should and must be overcome: "Minds must be forced to receive the principles on which they [the moral sciences] are established," Condorcet insisted zealously. A just and stable society depended upon the enlightenment of its citizens, and the learned had a duty to direct this enlightenment. "Dream," Condorcet commanded, "that enlightenment makes virtue easy, that the love of the common good and even the courage to devote oneself to it is, truly, the normal state of the enlightened man. In an ignorant man, justice is but a passion incompatible, perhaps, with gentleness. In an educated man justice is only humanity itself, submitted to the laws of reason."[47]

The French Academy received Condorcet's discourse poorly. Had he been alive, Voltaire might have warned him that "the obligation to speak, the awkwardness of having nothing to say, and the desire to be brilliant are three things calculated to make even the greatest man ridiculous"—but, of course, Condorcet did have something to say, which only made it worse.[48] Grimm reported that Condorcet's discourse was "without warmth, without harmony, without elegance, full of hackneyed ideas, and of a false and affected metaphysic."[49] However poorly conceived or delivered, the speech nevertheless revealed not only the persistence of Condorcet's commitment to the idea of philosophic struggle but also his growing interest in social science and public education as key elements in the establishment of a stable, just society. His first serious intellectual contribution to this ideal would be the development of what he termed *social mathematics*.

Although Condorcet's interest in social mathematics dates at least to 1772, his first serious attempt to apply it to the social and political realms did not arrive until the mid-1780s.[50] The first major result of his work was the difficultly titled *Essai sur l'application de l'analyse à la probabilité des décisions rendues à la pluralité des voix* (Essay on the Application of Analysis to the Probability of Decisions Rendered by a Plurality of Votes), published in 1785 (henceforth *Essay*).[51] Philosophically and ideologically, the work was linked to his *Vie de Turgot* (1786), an intellectual biography in which Condorcet sought to explain Turgot's policies and the principles behind them and to attribute an ultimately republican goal to Turgot's notion of legal despotism. The principles that guided Turgot's actions as a man of state, Condorcet demurred, provided the foundations for the *Essay*, which was in effect only an attempt to work out Turgot's ideas in a systematic fashion. Condorcet wrote in the introduction:

> A great man whose lessons, examples, and above all friendship I miss every day, was persuaded that the truths of the moral and political sciences are susceptible to the same certitude as that which forms the system of the physical sciences, even those branches that . . . seem to approach a mathematical certitude. . . . This opinion was dear to him because it led to the consoling hope that the human species will necessarily progress toward happiness and perfection, as it has so done in the knowledge of truth.[52]

Scholars have closely studied Condorcet's conclusions in the *Essay on Probability*.[53] Some of his ideas have been disproved, ridiculed, or ignored. Others have become common parlance in voting theory, particularly the so-called Condorcet paradox (or Condorcet winner). In its historical context, however, the *Essay* marked a critical step in the development of Condorcet's theory of education—a theory closely tied in his thought to the establishment of a feasible republican constitution. His mathematical analysis of voting procedures in the *Essay* implied the necessity of a high degree of education for all members of a voting body so that the probability of that body's decisions would be inclined toward the best or correct results. Such a conclusion in itself does not suggest a republican constitution is the best, but Condorcet went on to conclude that the greater the number of educated voters, the closer the probability of decisions would approach a moral certainty. A large, educated voting body would therefore be more just than a small one.

Condorcet worked out his theory using the example of a jury voting on the guilt or innocence of a defendant in a criminal case, although the theory works for any instance of voting in which one assumes a best or correct outcome is possible (which for Condorcet would have been all cases, since he argued that the goal of decision-making is to arrive at the most probable truth, not merely at a consensus). Suppose, he argued, that the probability of each juror's vote conforming to the truth (that is, whether the accused is truly guilty) is greater than one-half, or .5. Voting by a greater number of jurors would increase the probability, up to a moral certitude, that the decision would be correct. Mathematically, the probability would approach closer and closer to 1. If, on the contrary, the probability of each juror's correct judgment were less than .5, the probability that the votes would determine the right or best outcome would diminish according to the number of voters until one could be morally certain of an incorrect outcome. If the probability of each vote were exactly .5, of course, the probability of a correct decision would always be 50 percent—that is, no better than flipping a coin.[54] Obviously, in any such system of voting the desire would be to raise each juror's probability of correctness above .5. This, Condorcet realized quickly, was a serious problem in his society.

"This conclusion conducts us first," he wrote, "to a rather important point. A large assembly could not be composed of very enlightened men." Since society was largely uneducated or poorly educated, at least according to Condorcet, the probability of any vote in a general assembly would be below .5—and the larger the assembly, the greater the probability that its decisions would be bad or erroneous. "Since prejudice and ignorance about very important issues can exist," Condorcet noted, "one sees that it could be very dangerous to give a democratic constitution to an unenlightened people. A pure democracy would not be suitable except for a people much more enlightened, much more exempt from prejudice than any known in history."[55]

On the one hand, Condorcet seems to conclude that democracy, at least in the present state of society, was a very bad idea; the only practical alternative was an enlightened absolute monarchy—the traditional ideological position of the *parti philosophique*, pointing back to its Neostoic roots. In fact, he suggested this explicitly to Frederick of Prussia in May while discussing the conclusions of his *Essay*: "One must conclude that the happiness of peoples depends more on the enlightenment of those who govern them than on the form of political constitutions . . . that finally there is more hope in a monarchy than in a republic to see the destruction of abuses carried out promptly and in a tranquil manner."[56] (Was Condorcet merely pandering to Frederick, who fancied himself an enlightened man?)

On the other hand, Condorcet insisted in *Vie de Turgot*, published only a year later, that "a republican constitution is the best of all. It is the one in which the rights of man are conserved, since that of exercising legislative power, either by himself, or by his representatives, is one of these rights."[57] Even here, Condorcet equivocated, however. A republican constitution might be the best, he argued, but any political system not based on enlightenment would lack justice, since laws must not be an expression of "the arbitrary will of the majority," but rather of "truths deduced by reason from the principles of natural right."[58] Therefore, an enlightened monarchy must be more just than an unenlightened republic.

But did Condorcet really believe any longer that monarchy was a viable solution? Was he vacillating indecisively between support for absolute monarchy and representative republicanism? Was this the crux of his intellectual transformation? Had he, in fact, already adopted revolutionary ideas? It is hard to be definite, but we know that he openly proclaimed republican sentiments as early as 1789 and was one of the first in France to call for the abdication of the king in 1791, following the royal family's abortive flight to Varennes. But it is clear that Condorcet was already seeking philosophical and scientific justifications for his vision of modern republicanism by 1785. He found that justification most obviously in mathematics. According to his analysis, not only the greater probability of each vote but also the greater number of votes increased the certainty of a decision. "One has no legitimate motive," he concluded in the *Essay*, "for restraining the number of voters and submitting by that action the majority to the will of the smallest minority. Instead, in the case where one can form an assembly such that there is a great probability that its decisions will be true, there is a just motive for less enlightened men to submit their will to the decisions of this assembly."[59] In effect, eligibility for representative assemblies should be limited to those with a certain level of education, and popular voting should be evaluated proportionally, or weighed in favor of education, in order to achieve the best possible results—since Condorcet believed a proper education in the rational sciences and in critical thinking helped to guarantee voting independence and accuracy. "Calculus," he wrote in defense of his scheme, "becomes necessary as soon as the truth depends on exact and precise observations." In cases when someone's opinion can decide not only his own fate but also the fate of others, "it does not suffice in order to be just to believe in opinion. It is necessary to have [rational] motives for believing, and that these motives can be regarded as veritable proofs. Thus one ought not at all to regard as indifferent the means of evaluating, whenever possible, the degrees of probability which determine our decisions, and of assuring by this method the justice of our judgments and of our actions."[60]

So long as the educational level of society was low, a government dominated by an educated elite was the best and most just.[61] The key to establishing a functioning republic, Condorcet wrote in 1786, was obviously public instruction: "Having meditated for a long time on the means of ameliorating the fate of humanity, I have not been able to keep myself from believing that there is truly but one sole solution: it is to accelerate the progress of enlightenment. All other means have only a temporary and limited effect."[62] Eventually, Condorcet would conclude that the condition of eligibility for voting and the holding of office would be a diploma from the national educational system he proposed. Since this education would be required of every citizen, male and female, anyone who completed his or her studies would be eligible for office and would rank equally with other voters. We could hardly define such a political system as anything other than a representative democracy.[63]

The *Essay* was Condorcet's first original meditation on the conditions of possibility for an acceptable alternative to absolute monarchy.[64] As the political crises of the late 1780s deepened, so, too, did Condorcet's judgment that his "enlightened republic" (we might call it) was the best solution. His open proclamation of republican ideals in 1789 (with the pamphlet *Sentiments d'un républicain*) and public declaration of republican convictions in 1791 (in *De la république, ou un roi est-il nécessaire à la conservation de la liberté?*) resulted from the systematic development of his own philosophical views supported by his mathematical work. Condorcet's political ideas were, thus, philosophically independent of many other revolutionary ideas of the time—the result of which was to isolate him ideologically during the Revolution.

The political crises of the late 1780s gave Condorcet the much-awaited opportunity to promote his ideas in the public arena, first as a political pundit, and finally as a political representative.[65] He did not at all confine the implications of the *Essay* to the realm of philosophical conjecture, but he eagerly sought their application in the real world. His plan for public instruction in 1792, as well as his work on the Constitutional Committee in 1793, constituted the ultimate result of his political efforts.

By 1785, and probably earlier, Condorcet had moved decidedly toward open republican convictions. In the later 1780s, his association with

Americans may have helped him to overcome any lingering philosophical reservations about the feasibility of a republican constitution. He associated regularly with the Americans in Paris, especially the Anglo-American radical, Thomas Paine, and he was named an honorary citizen of New Haven, Connecticut, in 1787. After his marriage to Sophie de Grouchy at the end of 1785, their apartments at the *hôtel de la monnaie* became a prerevolutionary salon frequented by Jefferson, Paine, and Americanophiles such as Beaumarchais, Sieyès, La Rochefoucauld, and Lafayette (who was a close friend and had been a witness at Condorcet and Sophie's wedding).

Interpreted according to Condorcet's lights, America, in fact, became a useful example of his own philosophy in action. Although bothered by the continuation of slavery in the United States (which, following Paine, Condorcet asserted was a vestige of colonial tyranny, not a true expression of American ideals), Condorcet proclaimed American society to be a perfect illustration of an enlightened republic founded on education and equal opportunity. In *De l'influence de la révolution d'Amerique sur l'Europe* (1786), he asserted that the American Revolution provided four important lessons for Europe: it demonstrated that the rational reform of laws and customs was possible, that humanity could progress, that aggression against free men was futile, and that free trade was the answer to economic stagnation.[66] Philosophers had long recognized the natural rights of man, Condorcet affirmed, but "it does not suffice that they are written in the books of philosophers and in the hearts of virtuous men." What humanity needed was an example to follow. "America has given us this example," he continued. "The act by which it has declared its independence is a simple and sublime display of these rights, so sacred and forgotten for so long."[67] Moreover, and perhaps most importantly, as American society expanded into the vast North American continent, it would settle there an exponentially increasing number of enlightened people, free from the old prejudices that still plagued Europe. In a few generations, Condorcet foresaw, America might so contribute to the sciences that human progress would double. Having broken with the old world and its decadent customs and practices in order to establish a society based on reason and right, America was in the enviable position to save mankind. Universal education was the key, Condorcet concluded: "Let men be enlightened and soon you will see born the good without effort from the common will."[68]

The year 1786, then, saw Condorcet explicitly make his post-Turgot philosophical pivot in which he abandoned absolutism and adopted a modern republicanism because he had come to believe that reason could operate independently of political force—through its own self-evident truth—so long as education could be distributed widely enough.

After the dissolution of the Assembly of Notables in 1787, Condorcet engaged in further political pamphleteering promoting now recognizably

Condorcetienne political opinions that stemmed from his adaptation of the idea of the sciences as morally enlightening. In autumn 1788 and early 1789, he argued in *Sur la forme des élections* and in *Examen sur cette question: Est-il utile de diviser une assemblée nationale en plusieurs chambres?* that the Estates-General, and later the National Assembly, must represent the citizenry equally, according to the principles of political calculus he had devised. "It is absurd to suppose that a man would submit to execute the will of another man," he wrote, echoing the *Essay*, except when there is "the conviction that the opinion of this other man will conform to truth, to reason. . . . The goal that ought to be proposed in the constitution of such an assembly is to obtain true decisions."[69]

Condorcet sought election to the Estates-General while proclaiming these ideals. In various "Lettres d'un Gentilhomme à Messieurs du Tiers État," he insisted that he recognized no legitimate estates in the French nation, and therefore he desired to represent the Third Estate not as a nobleman but as a citizen. Perhaps with the unintentionally patronizing attitude of a highly educated aristocrat, he stated that his only distinction over many in the Third Estate was education, which made him better able to represent them well.[70] Despite apparent sincerity, Condorcet proved unacceptable to the Third Estate, however, and he did not receive their nomination. At the last minute he ran for the Second Estate, offering to represent the region of Mantes, where his family held a parcel of land. Unwilling to change his rhetoric, he gave a preelection speech before the Second Estate on March 10 decrying noble privilege. Needless to say, the Second Estate did not elect him.[71] "The nobles found me too populist," he wrote to Amélie Suard afterward. "And the non-nobles too moderate. Apparently because I am reasonable. I had against me in our assembly the aristocrats, the parlementaires, the landlords, the *dévots*, and half the slave traders. You judge if I could have succeeded."[72]

In Paris, the hotbed of more radical views, his chances were better. In September he was elected to the Paris Commune, where he worked with a new ally, Brissot. He also joined the new National Guard (although tradition has him carrying an umbrella in place of a sword while on duty) as well as various revolutionary clubs, including the Society of 1789.[73] He also continued to promote a close relationship between the state and the scientific community by representing the interests of the Academy of Sciences to the National Assembly.[74] Not until September 1791, however, did he win a place in a national body—the newly formed Legislative Assembly.

By then Condorcet's own personal revolution had begun. After the king and queen's failed attempt to flee France in June 1791, Condorcet openly declared himself a republican. On July 12 during a political meeting in the Cirque du Palais-Royal, he argued that since the king had abandoned his post he had consequently abdicated his authority, and the nation was free to found a republic.[75] The majority in the National Assembly maintained its commitment to a constitutional monarchy, however, and the constitutional monarchists, now including Lafayette, denounced and slandered Condorcet in the papers for expressing his radical opinions.[76] In fury, Condorcet wrote to Lafayette demanding an explanation: "For twelve years you have been counted among the defenders of liberty. If you do not change your conduct, in a few days you will be counted among its oppressors." That evening the National Guard, under the command of Lafayette and under orders from Bailly, then mayor of Paris, fired into a crowd on the Champ de Mars that was protesting against the monarchy.[77] Condorcet's wife and young daughter, Eliza, were present. Although they were unharmed, the event marked Condorcet's decisive break with Lafayette. He later claimed never to have spoken to Lafayette again.[78]

Such was the context in which Condorcet, now a confirmed republican, assumed office in the Legislative Assembly in September. In order to institute his ideas, he sought the presidency of the Committee on Public Instruction and began work on his education plan, described in detail in the *Cinq mémoires sur l'instruction publique*. As I indicated at the beginning of this chapter, Condorcet first put his committee's plan before the Legislative Assembly on April 20, 1792. Unfortunately, the assembly was in no mood that day to hear learned discourses on the long-term future of the nation. France was preparing to declare preemptive war on the Habsburg Empire, which in 1791 had signed the Declaration of Pilnitz, calling for the restoration of the Old Regime. In the middle of his report, Condorcet was obliged to stop speaking and step down from the rostrum so Marshal Dumouriez could read a gloomy report on the crisis, followed, much to Condorcet's expressed disgust, by Louis XVI's proposal for war against the king of Hungary and Bohemia.[79] To everyone's surprise, Condorcet unexpectedly returned to the rostrum afterward in order to insist that war had not been declared by a king against another king, but by a free nation against tyranny. Condorcet's education report, however, was quickly shelved in the excitement.[80] As we know, when Condorcet submitted his plan again in December in the newly formed Convention, almost none of the deputies were in a mood to hear it.

Specifically, Condorcet proposed a national system of public instruction designed to cultivate the talents of the whole population and to provide the foundation for a more democratic society that would promote what he considered the "progress of the human spirit." This meant that schools at all levels (the committee decided on five) would provide free instruction to all citizens, and that teachers should be paid out of public funds. The idea of free public instruction was itself practically unheard of (Talleyrand had only proposed free elementary education), but this was only the beginning of the plan's novelty. Condorcet also insisted that instruction should not be limited to men, since "women have the same rights as men; they have therefore the right to the same opportunity to acquire the enlightenment that alone can give them the means of truly exercising those rights with the same independence as men and to an equal extent."[81]

Universal instruction for all citizens, regardless of sex, was, in Condorcet's view, necessary not only for the establishment of political liberty but also for social, scientific, and technological improvement. When public instruction was open to all, Condorcet argued, society would thereby exponentially multiply the number of citizens "whose impartiality, disinterest, and enlightenment" would direct public opinion toward peace and the improvement of its institutions. Although not everyone was born with the same talents or could achieve the same level of learning, nevertheless, "an education that accustoms a people to sense the value of truth, to esteem those who discover it or who know how to employ it, is the only means of assuring the happiness and liberty of a people."[82]

The primary charge against Condorcet's plan in the Legislative Assembly debates, as indicated above, was that it was elitist and would somehow create an aristocracy of the learned. The plan did indeed promote the formation of a natural elite of "talent" and "genius." Public instruction, Condorcet insisted, must not "allow any talent to escape unnoticed," and it must offer such talent "all the aid reserved up to now to the children of the rich."[83] During the brief December debates in the Convention, some deputies also opposed Condorcet's belief in the virtue of instruction, arguing the Rousseauist doctrine that the arts and sciences in fact corrupt humanity from its natural goodness. French citizens, they argued, only had such need of education as could lead them to good republican virtues—that is, to a simple love of country and an unreflective, uncritical respect for the laws.[84] To Condorcet, such a doctrine revealed the perverse desire of those in power to keep the people ignorant and thus more easily ruled.[85] The inequality of talent and genius between people is real, he insisted, and not in the realm of the law to regulate. Rather, the business of the state is to make certain that all citizens are instructed such that they can exercise those rights guaranteed them by law and carry out their required duties (including voting and the holding of public office) independently of any other authority than their own proper reason. When everyone is instructed well, an inequality of talent "does not submit one person to another . . . it is neither an evil nor an injustice; and certainly it would be a fatal love of equality that feared to expand the class of the enlightened and augment its learning."[86]

It was also wrong, Condorcet argued, to introduce into public instruction anything other than the teaching of positive knowledge. Since critical and scientific reasoning was the source of liberty and the guide to virtue, Condorcet insisted that doctrine and dogma—whether spiritual or secular—had no place in public instruction. Distinguishing between "education" and "instruction," he argued that the duty of instruction was to teach truths of fact and mathematics and the principles of free, critical thinking. Education, however, taught political, moral, and religious opinions as unquestionable truths.[87] By teaching people what to believe, such an education enslaved minds and violated the principles of justice: "He who enters into society carrying with him the opinions that his education has given him is no longer free; he is the slave of his masters, and his irons are much more difficult to break."[88] Since truth alone was the foundation of prosperity, stability, and happiness, education ought not to consecrate established opinions, even to combat other errors, but must submit all opinions to free examination. This included the laws and constitution of the nation:

> But if one means that it is necessary to teach it [the constitution] as a doctrine conforming to the principles of universal reason, or to excite in its favor a blind enthusiasm that makes the citizens incapable of judging it; if one says to them: This is what you must adore and believe, then it is a kind of political religion that one wants to create; it is a chain that one prepares for minds, and one violates the most sacred rights of liberty under the pretext of teaching men to cherish it. The end of instruction is not to make men worship a perfect and complete legislation, but to make them capable of appreciating and correcting it.[89]

In other words, education teaches people what to think; instruction teaches them how to think. In order to prevent the system of public instruction from falling into the hands of political interests or extremists who would pervert its purpose, Condorcet proposed that "learned companies" [*compagnies savantes*] in each administrative department oversee instruction as well as scientific innovation at the highest level. At the head of these companies would be a National Society in Paris—clearly based on the old Academy of Sciences and representing an earlier model of collaboration between reason and power—to represent and direct all the rest. Each company would consist of the most meritorious experts in the fields of human knowledge and must choose its own members freely and independently, without political influence. Building such academies would be a simple matter, Condorcet explained, because, as he had long asserted, they already existed and had always existed. The enlightened associate naturally wherever they are, and need only be recognized officially by the state: "The public authority has no need to form learned societies; it recognizes them but does not create them. The Paris Academy of Sciences existed with Carcavi; the Society of London with Oldenburg; they were each an assembly of the most celebrated men of each nation, and they are still. Adopted by kings, they have continued to be what they were, and what they would have remained without them."[90]

As Condorcet had argued in 1779 during the mesmerism controversy, the importance of such learned societies was to make savants independent of unfounded opinion. Experts in their respective fields, their work judged only by their peers and according to its scientific merits, the savants of the learned societies were in the best position to direct public opinion. As knowledge and science progressed, it would become manifestly impossible for every citizen to acquire sufficient expertise to judge the merits of every new idea and new invention, or to choose what to add to the curriculum of public instruction. Citizens must be able to turn to experts they can trust, and "it is useful to them to have a sign by which they can recognize the opinion of learned men which, when it is unanimous and definitively formed, is nearly always in accord with the truth; and *voilà*, they will find this in a system of learned societies that embraces all the sciences and arts."[91]

Within three generations, Condorcet dared to predict, his system of free, universal public instruction would truly revolutionize society and achieve a general reformation of morals and politics. Indeed, one of the first achievements of education reform would be in politics. The politics of philosophy (founded on natural right and reason), Condorcet argued, would replace the politics of intrigue (founded only on petty self-interest) as the population began to understand where its true, rational, and common interests lay. Under such a system of public instruction, Condorcet believed, the philosophes would finally achieve their dream: the reform of the world according to the principles of scientific reason. In this brave new world, "the sole sovereign of a free people, the truth, of which men of genius are the ministers, will extend over the whole universe its sweet and irresistible power; from it all men will learn what they should desire for their happiness, and they will no longer desire but the common good of all."[92] As Condorcet prophesied in his posthumous *Esquisse*, such a system of instruction would so disseminate enlightenment that humanity would at last enter the Tenth Epoch of history.[93] Unfortunately for Condorcet, the Convention buried his plan, and he would not live to revive it.

In September 1792, Condorcet managed an appointment to the Constitutional Committee with, among others, Sieyès, Brissot, Thomas Paine, and Danton. Perhaps sensing that the writing of the constitution would be his last opportunity to influence the outcome of the Revolution, Condorcet so exhausted himself with the project that he could not complete his report to the Convention on February 15, 1793; other committee members took his place on the rostrum.[94] The constitution clearly embodied Condorcet's ambitions. It included a declaration of rights, abolition of the death penalty, provisions for public instruction following the outlines of Condorcet's rejected plan, and—in a direct assault on the Jacobins, who by then had assumed control of the Montagnard faction—a redrawing of the electoral map to deemphasize urban areas, the centers of Jacobin support.[95]

The Girondins supported the constitution—they had much to gain in a document that limited the power of the more radical Montagnard faction, dominated by the Jacobins. The Jacobins, of course, denounced the constitution and managed to delay further discussion of it. On May 29, the Committee of Public Safety formed a special Committee of Five to prepare a more acceptable constitution. On June 2 the Jacobins violently purged the Girondin leadership from the Convention, leaving the Committee of Five free to propose its constitution with little or no opposition.

Although in many ways an epitome of Condorcet's constitution (containing but 124, as opposed to 370, articles), the Jacobin constitution was fundamentally less democratic—primarily by establishing executive power in a council of twenty-four deputies appointed by the assembly, rather than in seven ministers directly elected by the people.[96] Such a system, of course, favored the dominant faction in the assembly. Incensed by this final outrage and the defeat of all his plans, Condorcet furiously took up the pen to denounce publicly the new constitution and accuse its writers of preparing the way for the return of the monarchy.[97] By this open act of opposition, Condorcet sided fatally with the recently crushed Girondins. On July 8 the Committee of General Security accused Condorcet of treason and conspiracy against the state and called for his arrest. Forewarned, he went into hiding. Eight months later he was captured in the southern suburbs of Paris and died in prison.[98]

NOTES

1. See Bill Readings, *The University in Ruins*, and, of course, Immanuel Kant, "An Answer to the Question: What Is Enlightenment?" (1784).
2. Condorcet to Voltaire, 1776, *Oeuvres de Condorcet*, 1: 113. See also chapter 4.
3. See Wright, *A Classical Republican in Eighteenth-Century France*, 204–5.
4. See "Rapport et projet de décret sur l'organization générale de l'instruction publique, présentés à l'Assemblée Nationale, au nom du comité d'instruction publique, les 20 et 21 Avril 1792," *Oeuvres de Condorcet*, 7: 449–573.
5. "Rapport et projet," 7: 450.
6. See Iain McLean, *Condorcet: Foundations of Social Choice and Political Theory*, introduction, 3–82, and Keith Baker, *Condorcet*.
7. "Propositions morales," in *Avertisements insérés par Condorcet dans l'édition complète des Oeuvres de Voltaire*, 1789, *Oeuvres de Condorcet*, 4: 539–40. "L'idee de la justice, du droit, se forme nécessairement de la même manière, dans tous les êtres sensibles capables des combinasions nécessaires pour acquérir ces idées. Elles seront donc uniformes. . . . [P]uisque tout être raisonnant juste sera conduit aux mêmes idées en morale comme en géométrie, il n'en est pas moins vrai que ces idées ne sont point arbitraires, mais certaines et invariables. Elles sont, en effet, la suite nécessaire des propriétés des êtres sensible et capable de raisonner; elles dérivent de leur nature." See also Turgot to Condorcet, December 28, 1773, *Oeuvres de Condorcet*, 1: 227. "Je vous dirai seulement que je ne crois pas que la morale en elle-même puisse être jamais locale. Ses principes sont partout fondés sur la nature de l'homme et sur ses rapports avec ses semblables, qui ne varient point, si ce n'est dans des circonstances très-extraordinaires."
8. *Sorberiana*, 163–64.
9. See *Esquisse*, tenth epoch, 274–94.
10. See R. R. Palmer, *The Improvement of Humanity*, Paul Hanson, *The Jacobin Republic under Fire*, Jay Smith, *Nobility Reimagined*, and also Lynn Hunt, *The Family Romance of the French Revolution*.

11. In the case of French history, see, for example, Eugen Weber, *Peasants into Frenchmen.*

12. First presented to the Legislative Assembly in April 1792.

13. See, in particular, R. R. Palmer, *The Improvement of Humanity*, 121–38.

14. J. Guillaume, *Procès-verbaux du comité d'instruction publique de la Convention Nationale* (Paris: 1891), 1: 121. Future Girondins who opposed or criticized the plan included Rabaut Saint-Étienne and J. F. Ducos.

15. M. Robespierre, *Oeuvres complètes de Robespierre, vol. 5, Les journaux . . . édition critique préparée par Gustav Laurent*, 207–11.

16. Not until late May 1793 did education reform return to the fore of political events as the Committee of Public Safety ordered a new Committee of Public Instruction, dominated by the abbé Sieyès, P. D. Daunou, and Joseph Lakanal, to draft a new plan. Their plan, presented to the Convention on June 26, attempted to incorporate the "moral and civic training" that many had found lacking in Condorcet's plan, but the new plan was also accused of "elitism" and in the end met the same fate as the Condorcet plan.

17. This idea is proved, in Palmer's view, by the partial revival of Condorcet's plan in 1795 by the *Ideologues*, who are judged to have been no great friends to democracy.

18. The notion that education should form citizenship and reverence for *la patrie* emerged in Montesquieu's *De l'esprit des lois*, Duclos's *Considérations sur les moeurs de ce siècle* (1750), and the abbé Coyer's *Dissertation sur le vieux mot de patrie* (1755). Keith M. Baker, *Condorcet*, 285–90.

19. See Smith, *Nobility Reimagined*. The Seven Years' War took place in the years 1756–1763.

20. Palmer, *Improvement of Humanity*, 94–101.

21. "Rapport et projet de décret sur l'organization générale de l'instruction publique, présentés à l'Assemblée Nationale, au nom du comité d'instruction publique, les 20 et 21 Avril 1792." *Oeuvres de Condorcet*, 7: 449–573.

22. On the ultimate fate and influence on Condorcet's plan, see Palmer, *Improvement of Humanity*; C. Kintzler, *Condorcet: l'Instruction publique et la naissance du citoyen*; Vial Francisque, *Condorcet et l'éducation démocratique*; Charles Coutel, *À l'école du Condorcet: contre l'Orléanisme des Esprits*; Henri Bigot, *Les idées de Condorcet sur l'instruction publique*; and Millard Hansen, "Condorcet's Liberal Philosophy of Education."

23. See Baker, *Condorcet*, 286–87. The notion that education should form citizenship and reverence for *la patrie* emerged in Montesquieu's *De l'esprit des lois*, Duclos's *Considérations sur les moeurs de ce siècle* (1750), and the abbé Coyer's *Dissertation sur le vieux mot de patrie* (1755).

24. *Vie de Turgot, Oeuvres de Condorcet*, 5: 1–233.

25. Turgot ordered du Pont de Nemours to outline his ministry's education plan in the *Mémoire sur les municipalités* (1775).

26. See, curiously, Jaucourt's article on "Desespoir" and its psychological effects in the *Encyclopédie* (4:882): "Cette triste passion qui nous trouble & qui nous fait perdre toute espérance, agit différemment dans l'esprit des hommes: quelquefois elle produit l'indolence & le repos; la nature accablée succombe sous la violence de la douleur: quelquefois en se privant des seules ressources qui lui restoient pour remedes, elle se fâche contre elle-même, & exige de soi la peine de son malheur, si l'on peut parler ainsi; alors, comme dit Charron, cette passion nous rend semblables aux petits enfans, qui par dépit de ce qu'on leur ôte un de leurs joüets, jettent les autres dans le feu. Quelquefois au contraire le desespoir produit les actions les plus hardies, redouble le courage, & fait sortir des plus grands périls."

27. See Condorcet's letter of resignation, October 1776, *Oeuvres de Condorcet*, 1: 297–98. "J'ai prononcé trop hautement mon opinion sur les ouvrages de M. Necker et sur sa personne, pour que je puisse garder une place qui dépendrait de lui. Je serais fâché d'être dépouillé, et encore plus d'être épargné par un homme dont j'ai dit ce que j'ai dit de M. Necker. Daignez donc, je vous prie, M. le Comte, si vous conservez encore quelque bonté pour moi, me faire savoir si je dois dépendre du département de M. Necker, et je vour supplierai dans ce cas de vouloir bien permettre que ce soit entre vos mains que je remettre ma démission."

28. See d'Alembert to Condorcet, September 28, 1777, and Condorcet to Turgot, September 1777, *Correspondance de Condorcet et de Turgot*, 297–98.

29. That is, for his *Lettres d'un théologien* (1774).

30. See Voltaire to Condorcet, April 9, 1777, *Oeuvres de Condorcet*, 1: 149: "La nature m'a joué un mauvais tour, mon très-respectable philosophe. Je combats un peu contre elle, mais je serais bientôt vancu. Je veux que vous me promettiez pour ma consolation de daigner prendre ma place à l'académie des paroles, quoique vous soyez le soutien de l'académie des choses, et d'être reçu par M. D'Alembert. J'irai me présenter là-haut ou là-bas, ou nulle part avec plus de confiance." On Condorcet's intractability, see Condorcet to Turgot, September 1777, *Correspondance de Condorcet et de Turgot*, 297. "Je regarde M. de La Vrillière non loué comme une bataille gagnée."

31. See d'Alembert to Voltaire, November 18, 1777, *Oeuvres Complètes de Voltaire*, 50: 312–13.

32. For example, Benjamin Franklin, to whom Condorcet sent a copy of the eulogy, noted, "I have read with highest pleasure your excellent Éloge de M. l'Hôpital. I knew you before as a great mathematician; I now consider you as one of the first among the politicians of Europe." *The Papers of Benjamin Franklin*, 25: 62. Cited in Badinter and Badinter, *Condorcet*, 165. Condorcet met Franklin in late 1776 or early 1777. He was on good terms with most of the American diplomats, including Jefferson, in Paris during the revolutionary era.

33. "Eloge de l'Hôpital," *Oeuvres de Condorcet*, 3: 463–566. Original French for full quotations, respectively: "Éclairé par son expérience, par l'étude, par de longues réflexions, pouvait-il ne pas sentir combien il faut se hâter de détruire ces abus qui attaquent sourdement les moeurs publiques et le caractère national, enfantent les vices, détruisent les vertus . . . ?" "Au milieu du plus violent fanatisme, il fit entendre la voix de la raison et de l'humanité; au sein de l'anarchie et de la révolte, il défendit avec un courage égal, et l'autorité du roi et les droits de la nation"; "la corruption de son siècle, les intrigues de la cour n'altérèrent ni son intégrité, ni sa franchise." "Mais nous ne le plaignons point d'avoir obtenu la haine des ennemies du peuple; pour une âme forte, cette haine est un bien: c'est la preuve la plus frappante qu'on a servi la patrie"; "le chancelier ne doit jamais perdre de vue que les magistrats ont été institués pour le peuple, et que, placé à leur tête, il leur doit, non de défendre leurs prétentions, mais de leur assurer la liberté de remplir leurs devoirs. Si la crainte, la bassesse, l'avidité, la partialité corrompent la pureté des jugements . . . si les magistrats s'abaissent jusqu'à se rendre les instruments des passions des hommes puissants, ou les complices de leurs intrigues; s'ils négligent leurs fonctions utiles, pour aspirer à un simulacre de pouvoir qu'ils ne peuvent obtenir qu'aux dépens de la prosperité publique: qu'alors ils trouvent dans leur chef, un censeur plus occupé de les éclairer que de les punir."

34. See, for example, Condorcet's eulogy of Røemer 1772. *Oeuvres de Condorcet*, 2: 76–82.

35. *Oeuvres de Condorcet*, 2: 355. "S'il abuse de son pouvoir, il apprend bientôt, par la résistance de ses sujets même, que les rois ont été établis pour les peuples, et non les peuples pour les rois, et qu'il n'a reçu l'empire sur la nature que pour servir à conserver dans l'univers l'ordre que la Providence y a établi."

36. On Condorcet and the mesmerism controversy, see Robert Darnton, *Mesmerism and the End of the Enlightenment in France*; Baker, *Condorcet*, 99–104; and Robinet, *Condorcet, sa vie, son oeuvre*, 21.

37. Condorcet also worked in 1784 with Franklin and Bailly to prepare a report on mesmerism, ultimately rejecting its scientific claims. See Darnton, *Mesmerism*, 61–64, and Condorcet, "Raisons qui m'ont empêché jusqu'ici de croire au magnétisme animal," Bibliothèque de l'Institut de France, MS 833, ff. 231–47 (reproduced in Darnton, *Mesmerism*, 189–92).

38. J.-P. Brissot de Warville, *De la vérité*, 165–66. See Darnton, *Mesmerism,* 90–98, and Hahn, *The Anatomy of Scientific Institution*, 150–54.

39. Bibliothèque de l'Institut de France, MS 876, ff. 95–96. Cited in Baker, *Condorcet*, 102: "Les académies ont deux utilités incontestibles: La première d'être une barrière toujours opposé au charlatanisme dans tous les genres et c'est pour cela que tant de gens s'en plaignent; la deuxième de maintenir dans les sciences les bonnes méthodes et d'empêcher aucune branche des sciences d'être absolument abandonnée. Elles en ont une troisième très importante tant que les savants ne dédaigneront l'opinion populaire, c'est de les en rendre indépendants. Un chimiste, un anatomiste, un géomètre membre d'une académie n'a pas besoin de faire des tours de

charlatans pour jouir auprès des ignorants de la réputation d'un savant. C'est ensuite à ses ouvrages à lui mériter de la célébrité ou de la gloire."

40. See Badinter and Badinter, *Condorcet*, 198–200.

41. See Voltaire, *Letters on England*, article "Academies." "All I can make out . . . in these fine discourses is that the member elect, having affirmed that his predecessor was a great man, the Cardinal de Richelieu was a very great man, Chancellor Séguier a fairly great man, Louis XIV more than a great man, the Director replies in the same way that the member elect might also be a sort of great man, and that as for himself, the Director, he does not decline his own share."

42. *Oeuvres de Condorcet*, 1: 397. "Voyez, maintenant, d'un bout de l'Europe à l'autre, les hommes éclairés réunir tous leur efforts pour le bien de l'humanité, et tourner vers cet objet seul toutes leurs forces avec un courage et un concert dont aucun siècle n'a donné l'exemple."

43. *Oeuvres de Condorcet*, 1: 390. "Ces vérités premières, ces méthodes répandues chez toutes les nations ne peuvent plus s'anéantir; le genre humain ne reverra plus ces alternatives d'obscurité et de lumière auxquelles on a cru longtemps que la nature l'avait éternellement condamné. Il n'est plus au pouvoir des hommes d'éteindre le flambeau allumé par le génie; et une révolution dans le globe pourrait seule y ramener les ténèbres."

44. Of course, Samuel Sorbière, for one, had said this more than a century earlier. See chapter 1.

45. *Oeuvres de Condorcet*, 1: 392. "Tout serait égal entre elles pour un être qui, étranger à notre espèce, étudierait la société humaine comme nous étudions celle des castors ou des abeilles"; "on ne peut, en effet, s'empêcher de voir qu'appuyées comme les sciences physiques sur l'observation des faits, elles doivent suivre la même méthod, acquérir une langue également exacte et précise, atteindre au même degré de certitude."

46. See, especially, David Bates's discussion of Condorcet's analysis of the political decision in *Enlightenment Aberrations*, 78–97.

47. *Oeuvres de Condorcet*, 1: 393, 395. "Les principes sur lesqueles elles [moral sciences] sont établis ont besoin de forcer, pour ainsi dire, les esprits à les recevoir." "Songez que les lumières rendent les vertus faciles que l'amour de bien général, et même le courage de s'y dévouer, est, pour ainsi dire, l'état habituel de l'homme éclairé. Dans l'homme ignorant, la justice n'est qu'une passion incompatible peut-être avec la douceur; dans l'homme instruit, elle n'est que l'humanité même, soumise aux lois de la raison."

48. Voltaire, *Letters on England*, "Academies."

49. *Correspondance littéraire*, 3: 84. Cited in Badinter and Badinter, *Condorcet*, 201.

50. See letter from Condorcet to Turgot, September 3, 1772, in *Correspondance de Condorcet et de Turgot*, 97–98, and Condorcet to Suard, July–August 1772, XLVII in *Correspondance inédite de Condorcet et Mme Suard*, 87–88.

51. For the most astute analysis of this text in recent years, see Bates, *Enlightenment Aberrations*, 78–97.

52. Condorcet, *Essai sur l'application de l'analyse à la probabilité des décisions rendues à la pluralité des voix*, 1785, in *Condorcet sur les élections*, 9. "Un grand homme, dont je regretterai toujours les leçons, les exemples, et sur-tout l'amitié, étoit persuadé que les verités des Sciences morales et politiques, sont susceptibles de la même certitude que celles qui forment le système des Sciences physiques, et même que les branches de ces Sciences qui, comme l'Astronomie, paroissent approcher de la certitude mathèmatique. Cette opinion lui étoit chère, parce qu'elle conduit à l'esperance consolante que l'espèce humaine sera nécessairement des progrès vers le bonheur et la perfection, comme elle en a fait dans la connoissance de la verité."

53. See Ian McLean, *Condorcet*, Introduction, section 2, "The Theory of Voting," 32–48. See also Gilles-Gaston Granger, *La Mathématique sociale du Marquis de Condorcet*.

54. Condorcet, *Essai*, 29–30. Incidentally, Condorcet calculated a "moral certainty" rather precisely, through a rather convoluted set of calculations, arriving at the figure of 1/144,768, thus establishing the threshold of moral certainty at .999993092, roughly.

55. Condorcet, *Essai*, 29–30. "Cette conclusion conduit d'abord à une remarque assez importante. Une assemblée très-nombreuse ne peut pas être composée d'hommes très-éclairés." "Or, comme ces préjugés, cette ignorance, peuvent exister sur des objets très-importans . . . on

voit qu'il peut être dangereux de donner une constitution démocratique à un peuple sans lumières. une démocratie pure ne pourroit même convenir qu'à un peuple beaucoup plus éclairé, beaucoup plus exempt de préjugés qu'aucun de ceux que nous connoissons par l'Histoire."

56. Condorcet to Frederick of Prussia, May 2, 1785, *Oeuvres de Condorcet*, 1: 306. "Le second résultat est l'impossibilité de parvenir . . . à remplir les conditions qu'on doit exiger, à moins que ces décisions ne soient rendues par des hommes très-éclairés: d'où l'on doit conclure que le bonheur des peuples dépend plus des lumières de ceux qui les gouvernent que de la forme des constitutions politiques . . . qu'enfin il y a plus d'espérance dans une monarchie que dans une république, de voir la destruction des abus s'opérer avec promptitude et d'une manière tranquille."

57. *Vie de Turgot*, in *Oeuvres de Condorcet*, 5: 209–10. "Une constitution républicaine est la meilleure de toutes. C'est celle où tous les droits de l'homme sont conservés, puisque celui d'exercer le pouvoir législatif, soit par lui-même, soit par ses répresentants, est un de ces droits."

58. *Oeuvres de Condorcet*, 5: 211.

59. Condorcet, *Essai*, 31. "On n'a aucun motif légitime pour restreindre le nombre des Votans, et soumettre par-là le plus grand nombre à la volunté du plus petit: au lieu que dans le cas où l'on peut former une assemblée, telle qu'il y ait une très-grande probabilité que ses décisions seront vraies, il y a un motif juste pour les hommes moins élcairés que ses Membres, de soumettre leurs volontés aux décisions de cette assemblée."

60. *Essai*, 171–72: "Le Calcul devient nécessaire aussi-tôt que la vérité dépend d'observations exactes et précises"; "il ne suffit pas pour être juste de croire une opinion, il faut avoir de plus de motifs de la croire, et que ces motifs puissent être regardés comme de véritables preuves. Ainsi l'on ne doit point regarder comme indifférens les moyens d'évaluer, toutes les fois qu'il est possible, les degrés de la probabilité qui détermine nos décisions, et d'assurer par cette méthode la justice de nos jugemens et de nos actions."

61. On the perpetuation of similar ideals in the nineteenth century, see Alan Kahan, *Aristocratic Liberalism: The Social and Political Thought of Jacob Burckhardt, John Stuart Mill, and Alexis de Tocqueville.*

62. *De l'influence de la révolution d'Amerique sur l'Europe* (1786), in *Oeuvres de Condorcet*, 8: 30. "Occupé à méditer depuis longtemps sur les moyens d'améliorer le sort de l'humanité, je n'ai pu me défendre de croire qu'il n'y en a réellement qu'un seul: c'est d'accélérer le progrès des lumières. Tout autre moyen n'a qu'un effet passager et borné."

63. See the *Cinq mémoires*, 1st Mém, 77.

64. See also his "Tableau générale de la science, qui a pour objet l'application du calcul aux sciences politiques et morales," first published in the *Journal d'Instruction sociale*, 1795, but written earlier. *Oeuvres de Condorcet*, 1: 539–73.

65. A few of the traditional accounts of Condorcet's activities during the Revolution are Franck Alengry, *Condorcet: Guide de la Révolution Française*, and Léon Cahen, *Condorcet et la Révolution Française.*

66. Although written in 1786, and dedicated to Lafayette, the piece was published in 1788 in the pro-American journal of the Italian adventurer and American immigrant, Filippo Mazzei, the *Recherches historique et politiques sur les États-Unis*. Jefferson encouraged Mazzei's journal in an effort to counter anti-American propaganda in Europe; Condorcet inserted several pieces, and he helped edit the journal.

67. *Oeuvres de Condorcet*, 8: 11. "Mais il ne suffit pas qu'ils soient écrits dans les livres des philosophes et dans le coeur des hommes vertueux. . . . L'Amérique nous a donné cet example. L'acte qui a déclaré son indépendance est une exposition simple et sublime de ces droits si sacré et si longtemps oubliés."

68. *Oeuvres de Condorcet*, 8: 30. "Quand même on avouerait que des erreurs, des fables, des législations combinées, non d'après la raison, mais d'après les préjugés locaux, ont fait le bonheur de quelques nations, on serait forcé d'avouer aussi que partout, ce bien trop vanté a disparu en peu de temps, pour faire place à des maux que la raison n'a pas encore pu guérir après plusieurs siècles. Que les hommes soient éclairés, et bientôt vous verrez le bien naître, sans effort, de la volonté commune."

69. *Oeuvres de Condorcet*, 9: 334–35. "Il est absurde de supposer qu'un homme se soumette à exécuter la volunté d'un autre home . . . la conviction que l'opinion de cet autre homme sera conforme à la vérité, à la raison. . . . Le but qu'on doit se proposer, dans la constitution d'une telle assemblée, est donc d'obtenir des decisions vraies."

70. *Oeuvres de Condorcet*, 9: 215–19. Condorcet's campaign promises to the Third Estate were to ratify a declaration of rights, a constitution, and fair tax reform.

71. Archives Nationales, BIII 80, ff. 148–49, 338–40. Reproduced in Cahen, *Condorcet et la Révolution*, 101–8.

72. *Correspondance inédite de Condorcet et Suard*, 249–50. "Les nobles m'ont trouvé trop populaire, et les non-nobles trop modéré. Apparemment que je suis raisonnable. J'avais contre moi, dans notre assemblée, les aristocrats, les parlementaires, les propriétaires d'habitation, les dévots et la moitie des négriers. Jugez si je pouvais réussir."

73. See M. Norvins, *Essais sur la Révolution française*, and Badinter and Badinter, *Condorcet*, 308–10.

74. See, for example, his *Discours prononcé à l'Assemblée National, par M. de Condorcet, au nom de l'Académie des sciences, à la séance du 12 juin 1790*; *Lettre du secrétaire de l'Académie des sciences à M. le Président de l'Assemblée Nationale, 10 juin 1790*, and *Lettre de Condorcet à l'Assemblée Nationale* (as secretary of the Academy of Sciences), *Oeuvres de Condorcet*, 1: 508–11, 512–15, 525–26.

75. *De la République, ou un roi est-il nécessaire à la conservation de la liberté?* In *Oeuvres de Condorcet*, 12: 227–37.

76. See Badinter and Badinter, *Condorcet*, 382–83, citing the *Gazette de Paris*, July 17.

77. See Doyle, *French Revolution*, 153–55.

78. Condorcet reported this incident in 1794 in the "Fragment de Justification," written a few days before his death. *Oeuvres de Condorcet*, 1: 574–605.

79. The Emperor Leopold II died in March and his successor, Francis II, had not yet been formally elected emperor by mid-April.

80. Condorcet's remarkable speech attempted to legitimize the declaration of war as a function of the French Constitution, not as the orders of a monarch, and it excoriated the "feudal servitude and humiliating inequality" that would result from the restoration Austria sought. *Oeuvres de Condorcet*, 9: 443–55.

81. *Cinq Mémoires*, 100. "Enfin, les femmes ont les mêmes droits que les hommes; elles ont donc celui d'obtenir les mêmes facilités pour acquérir les lumières qui seules peuvent leur donner les moyens d'exercer réellement ces droits avec une même indépendance et dans une égale étendue."

82. *Cinq Mémoires*, 223. "Une éducation qui accoutume à sentir le prix de la vérité, à éstimer ceux qui la découvrent ou qui savent l'employer, est le seul moyen d'assurer la félicité et la liberté d'un peuple."

83. *Cinq Mémoires*, 68. "Laissât échapper aucun talent sans être aperçu, et qui offrît alors tous les secours réservés jusqu'ici aux enfants des riches."

84. J. Guillaume, *Procès-verbaux du comité d'instruction publique de la Convention Nationale*, 1: 124, 238, 278. See also Palmer, *Improvement*, 129–39.

85. *Cinq Mémoires*, 220.

86. *Cinq Mémoires*, 65. "Mais si cette inégalité ne soumet pas un homme à un autre, si elle offre un appui au plus faible, sans lui donner un maître, elle n'est ni un mal, ni une injustice; et certes, ce serait un amour de l'égalité bien funeste que celui qui craindrait d'étendre la classe des hommes éclairés et d'y augmenter les lumières."

87. Condorcet possibly reasoned such differences from the Latin origins of the terms: Education derives from *ex* + *ducere*, or "to lead out." Instruction derives from *in* + *struere*, or "to build on."

88. *Cinq Mémoires*, 85. "Celui qui en entrant dans la société y porte des opinions que son éducation lui a données n'est plus un homme libre; il est l'esclave de ses maîtres, et ses fers sont d'autant plus difficiles à rompre."

89. *Cinq Mémoires*, 93. "Mais si on entend qu'il faut l'enseigner comme une doctrine conforme qux principes de la raison universelle, ou exciter en sa faveur un aveugle enthousiasme qui rende les citoyens incapables de la juger; si on leur dit: Voilà ce que vous devez adorer

et croire, alors c'est une espèce de religion politique que l'on veut créer; c'est une chaîne que l'on prépare aux esprits, et on viole la liberté dans ses droits les plus sacrés, sous prétexte d'apprendre à la chérir. Le but de l'instruction n'est pas de faire admirer aux hommes une législation toute faite, mais de les rendre capables de l'apprécier et de la corriger."

90. *Cinq Mémoires*, 166. "Les sociétés savantes n'ont pas eu besoin de la puissance publique pour se former; elle les a reconnues et ne les a pas créées. L'Académie des sciences de Paris existait chez Caravaci; la socitété de Londres chez Oldenbourg; elles étaient l'une et l'autre l'assemblé des hommes les plus célèbres de chaque nation, et elles le sont encore. Adoptées par les rois, elles ont continué d'être ce qu'elles avaient été, ce qu'elles seraient restées sans eux."

91. *Cinq Mémoires*, 211–12. "Il leur utile d'avoir un signe auquel ils puissent reconnaître l'opinion des hommes éclairés, qui, lorsqu'elle est unanime et définitivement formée, se trouve presque toujours d'accord avec la vérité; et voilà ce qu'ils trouveront dans un système de sociétés qui embrasserait toutes les sciences et tous les arts."

92. *Cinq Mémoires*, 272. "Le seul soverain des peoples libres, la vérité, dont les hommes de génie sont les ministres, étendra sur l'univers entier sa douce et irrésistible puissance; par elle tous les hommes apprendront ce qu'ils doivent vouloir pour leur boneur, et ils ne voudront plus que le bien commun de tous."

93. See *Esquisse*, 274–94.

94. See Badinter and Badinter, *Condorcet*, 604–5.

95. See "Projet de Déclaration des Droits," *Oeuvres de Condorcet*, 12: 417–22, and Constitution, titre premier, *Oeuvres de Condorcet*, 12: 423–24.

96. Constitution, titre v, section première, article premier, *Oeuvres de Condorcet*, 12: 446.

97. "Aux citoyens français sur la nouvelle Constitution," *Oeuvres de Condorcet*, 12: 653–75.

98. Interesting documents concerning Condorcet's adventures as a fugitive are found in the Bibliothèque de l'Institut, MSS 848–49.

Epilogue

Despite endless debates, to many historians the French Revolution remains the birth of political modernity, unleashing onto the world the ideological struggles that characterize politics to this very day. In this world, as the revolutionaries seem to have understood well, education is intimately bound up with the exercise of power (why else do we still fight over issues such as evolution and creationism in the schools?). Condorcet's vision of eliminating patriotic indoctrination and national identity formation from national education could not free him from the fundamental truth that knowledge is power—in fact, he recognized this truth profoundly and sought, therefore, to diffuse power as widely as possible by making education the enabler of democracy. To actualize his plan, however, he had to exercise power himself—and to attempt to create an agency of savants who would then exercise tremendous political power by having absolute and unhindered authority over education. Like his friend and mentor, Turgot, Condorcet found the need for power unavoidable, since the moral idea of the sciences—the dream of the philosophes—was ultimately to unify power and reason. Condorcet was, as a result, inexorably drawn into the ideological and political battles of the Revolution—and eventually into the bitter struggle between the Girondins and the Montagnards in 1793, with tragic consequences.

With Condorcet's death, the Enlightenment of the *parti philosophique*, or at least the Voltairean faction of it, was over. Although he had sought to transform what amounted to a monarchist political ideology based on Neostoic principles that had emerged in the seventeenth century into a revolutionary liberal republican one based on those same principles, Condorcet did not have the support to implement his ideas, or even to protect himself from repression. The constitution he and his committee proposed was indeed more democratic than the subsequent Jacobin one, but by mid-1793 the Jacobin dictatorship had begun and would not end until the Thermidorian reaction of summer 1794. That was too late for Condorcet.

Ultimately, however, the line of Enlightenment thought Condorcet transformed from the seventeenth-century Neostoic idea of the sciences had great influence both on the technocratic political culture that began to emerge in post-Revolutionary France and on historical perceptions of the Enlightenment. Certainly with institutions such as the Grandes Écoles, France began to place greater value on the formation of an expert elite ruling class than on abstract ideas such as egalitarianism. As for historical perceptions of the Enlightenment, I do not think it is entirely out of bounds to suggest that the Enlightenment idea of education as the great emancipator has become a sort of "self-evident" truth—certainly in American culture, in which it would be voiced in almost the same terms by Benjamin Franklin in the eighteenth century, Frederick Douglass in the nineteenth century, and President Barack Obama in the twenty-first.[1] As I have tried to show in the preceding pages, the quintessential Enlightenment conception of education that Condorcet articulated so clearly was the result of a historical process that transformed and then applied Samuel Sorbière's Neostoic and highly elitist idea of the sciences as morally enlightening to a particular political and social situation that emerged from the trauma of 1776 and the failure of Turgot's reforms. Sorbière, on the eve of the foundation of the Academy of Sciences, expressed the idea that the sciences made the savant morally better; Fontenelle valorized—even deified—the savant as the natural moral elite of society; Turgot sought to legitimate absolute power through an appeal to the work of the savant; and Condorcet tried in vain to transform all of society in order to put the savant at the center. And the more elitist idea of the sciences survives quite well in the contemporary academic world, in which sociologists claim to see behind the veil of false consciousness to the true social forces that drive our actions; in which physicists claim to see beyond the world of simple sense perceptions to the subatomic nature of reality; and in which the microbiologists claim to see beyond the illusions of human free will to the underlying chemical processes that drive us unconsciously. Despite the anti-intellectual and envious mistrust leveled against those who can claim special knowledge of reality through disciplined scientific study, Americans in general—the children of the Enlightenment?—for the most part want to seize that special knowledge for their children so that they, too, might better understand the world and put that knowledge to use for their own freedom, profit, and interest.

And some of us still hope that, through the sciences, we might make the world better by coming to a more careful and disciplined understanding of both it and our place in it. That makes us truly the children of the Enlightenment—and perhaps subject to the same hubris as the philosophes.

NOTE

1. . See Benjamin Franklin to Condorcet, March 20, 1774, *The Papers of Benjamin Franklin*, William Willcox, ed., 151–52; Frederick Douglass, "The Blessings of Liberty and Education" (1894); Barack Obama, 2011 commencement address as Hampton University, http://www.whitehouse.gov/the-press-office/remarks-president-hampton-university-commencement.

Bibliography

PRIMARY SOURCES

Manuscripts

Archives Nationales, Paris.
BIII 80, ff. 148–49, 338–40

Bibliothèque de l'Arsenal, Paris.
MS 7464

Bibliothèque de l'Institut de France, Paris.
Condorcet MS 833, ff. 231–47
Condorcet MSS 848–49
Condorcet MS 876, ff. 95–96

Bibliothèque Nationale, Paris.
Manuscrits, *Cinq Cents de Colbert*, n° 485

Printed Primary Sources

Arc, Philippe Auguste de Sainte-Foy, chevalier d'. *La noblesse militaire, opposée à un Quatrième Ordre de l'Etat.* 1789, n.p.

Argenson, R. L. de Voyer de Paulmy, marquis d'. *Considérations sur le gouvernement ancien et présent de la France.* Amsterdam, 1765.

Aristotle. *Ethics of Aristotle.* Translated by J. A. K. Thomson. New York: Penguin Books, 1976.

Brissot de Warville, Jacques-Pierre. *De la vérité, ou méditations sur les moyens de parvenir à la vérité dans toutes les connaissances humaines.* Neuchâtel, 1782.

Charron, Pierre. *Toutes les Oeuvres de Pierre Charron.* Paris: Jacques Villery, 1635.

Condorcet, Jean Antoine Nicolas de Caritat, marquis de. *Cinq mémoires sur l'instruction publique (1791).* Paris: Flammarion, 1994.

———. *Condorcet sur les élections et autres textes.* Edited by Olivier de Bernon. Paris: Fayard, 1986.

———. *Correspondance inédite de Condorcet et de Turgot 1770–1779.* Edited by Charles Henry. Paris, 1880.

———. *Correspondance inédite de Condorcet et Mme Suard, M. Suard et Garat: 1771–1791*. Paris: Fayard, 1988.
———. *Esquisse d'un tableau historique des progrès de l'esprit humain*. Paris: Flammarion, 1988.
———. *Histoires et mémoires de l'académie royale des sciences, 1783*. Paris, 1786.
———. *Journal d'instruction sociale. Prospectus*. Paris, 1793.
———. *Journal de la société de 1789 (and Prospectus)*. 15 nos. Paris, June 5–September 15, 1790.
———. *Oeuvres de Condorcet*. Edited by Arthur Condorcet-O'Connor and François Arago. Paris, 1847–1849.
———. *Réflexions et notes sur l'éducation*. Naples: Bibliopolis, 1983.
———. *Réflexions sur l'esclavage des nègres*. Neuchatel, 1781.
———. *Vie de Voltaire*. Edited by Elisabeth Badinter. Paris: Quai Voltaire, 1994.
Condorcet, Jean Antoine Nicolas de Caritat, marquis de, Jules Michel Duhamel, and Emmanuel Joseph Sieyès, eds. *Journal d'instruction sociale*. 6 vols. Paris, June 1–6, 1793.
Constant, Benjamin. *De la liberté des anciens comparée à celle des modernes*. Paris, 1819.
Coyer, Gabriel-François. *La noblesse commerçante*. London, 1756.
D'Alembert, Jean le Rond. "Analyse de l'Esprit des Loix." In Montesquieu, *Oeuvres complètes de Montesquieu*. Vol. 1. Edited by M. André Masson. Paris: Nagel, 1950.
———. *Oeuvres complètes de d'Alembert*. Paris: Belin, 1821–1822.
Diderot, Denis, and Jean le Rond d'Alembert. *Encyclopédie ou dictionnaire raisonné des sciences, des arts et des métiers (articles choisis)*. Paris: GF-Flammarion, 1986.
Du Pont de Nemours, Pierre Samuel. *Mémoires sur la vie et les ouvrages de M. Turgot*. Philadelphia [Paris], 1782.
Fontenelle, Bernard le Bouvier de. *Conversations on the Plurality of Worlds*. Translated by H. A. Hargreaves. Berkeley: University of California Press, 1990.
———. *Oeuvres complètes*. Volumes 1–7. Paris: Fayard, 1993–1994.
———. *Oeuvres complètes de Fontenelle*. Edited by G.-B. Depping. Paris, 1818. Reprint, Geneva: Slatkine Reprints, 1968.
Franklin, Benjamin. *The Papers of Benjamin Franklin*. Edited by William Willcox. New Haven, CT: Yale University Press, 1976.
Gassendi, Pierre. *Peiresc, 1580–1637: Vie de l'illustre Nicolas-Claude Fabri de Peiresc, conseiller au parlement d'Aix*. Translated by Roger Lasalle. Paris: Belin, 1992.
———. *Syntagma Philosophiae Epicuri*. Lyon, 1658.
Grandjean de Fouchy, Jean-Paul. "Eloge de Fontenelle." In *Histoire de l'académie royale des sciences, de l'année M.D.CCLVII* [1762], *etc. Nouvelle Centurie*. Vol. 15. Amsterdam: J. Schreuder, 1769.
———. "Eloge de Mairan." In *Histoire de l'academie royale des sciences*, 1771. Paris, 1774.
———. *Eloges des académiciens de l'académie royale des sciences, morts depuis l'an 1744*. Vol. 1. Paris, 1761.
Grimm, Friedrich Melchior, et al. *Correspondance litteraire, philosophique, et critique*. Edited by Maurice Tourneux. 16 vols. Paris: Garnier, 1877–1882.
Guillaume, J. *Procès-verbaux du comité d'instruction publique de la Convention Nationale*. Paris, 1891.
Hobbes, Thomas. *De Cive*. Amsterdam, 1647.
———. *De Cive*. Translated by Howard Warrender. Oxford: Clarendon Press, 1983.
———. *Elémens philosophique du citoyen, où les fondemens de la société civile sont découverts par Th. Hobbes et traduits en français par un de ses amis*. Translated by Samuel Sorbière. Amsterdam: Blaeu, 1649.
———. *Le corps politique, ou les élemens de la loi morale et civile, avec des réflexions sur la loi de nature, sur les serments, sur les pactes et les diverses sortes de gouvernemens, leurs changemens et leurs relations, par Th. Hobbes*. Translated by Samuel Sorbière. Leiden: J. & D. Elzevier, 1653.
Huygens, Christiaan. *Oeuvres complètes*. 22 vols. The Hague: M. Nijhoff, 1888–1950.

Kant, Immanuel. "An Answer to the Question: What Is Enlightenment?" In *Kant: Political Writings*, edited and translated by H. S. Reiss. Cambridge: Cambridge University Press, 1970.

Lagrange, Joseph Louis, comte de. *Oeuvres de Lagrange*. Paris, 1892.

Louis XVI. *Mémoires pour l'instruction du Dauphin*. Paris: Imprimerie Nationale, 1992.

Marolles, Michel, l'abbé de. *Mémoires de Michel de Marolles abbé de Villeloin*. 3 vols. Amsterdam, 1755.

Montaigne, Michel de. *The Complete Essays*. Translated by M. A. Screech. New York: Penguin Books, 1991.

Montesquieu, Charles-Louis de Secondat, baron de. *Oeuvres complètes de Montesquieu*. Edited by M. André Masson. Paris: Nagel, 1950.

Reglements de la Société des Amis des Noirs. Paris, 1788.

Robespierre, Maximilien. *Oeuvres complètes de Robespierre*. Vol. 5, *Les journaux*. Edition critique préparée par Gustav Laurent. Gap: Louis-Jean, 1961.

Sorbière, Samuel. *Discours sceptique sur le passage du chyle, & le mouvement du Coeur* . . . Leiden: De J. Maire, 1648.

———. *Lettres et discours de M. de Sorbière sur diverses matières curieuses*. Paris, 1660.

———. *Relation d'un voyage en angleterre, où sont touchées plusieurs choses, qui regardent l'estat des sciences, et de la religion, et autres matieres curieuses*. Cologne: Pierre Michel, 1666.

———. *Relations, Lettres, et discours sur diverses matières curieuses*. Paris, 1660.

———. *Sorberiana, ou les pensées critiques de M. de Sorbiere, recueilliés par M. Graverol, avocat de la ville de Nismes. Avec des memoires pour la vie de Mess. De Sorbiere et Cortelier*. 2nd edition. Paris: Chez Florentin et Pierre Delaulne, 1695.

———. *Sorberiana, sive excerpta ex ore Samuelis Sorbière. Prodeunt ex musaeo Francisci Graverol, Accedunt ejusdem tum epistola de vita et scriptis Samuelis Sorbière et Joan.-Bapt. Cotelier, tum epulae ferales, Editio auctior et emendatior*. Tolosae [Toulouse]: G.-L. Colomyez, 1694.

Turgot, A. Robert-Jacques. *Oeuvres de Tugot*. Paris: Alcan, 1913–1923.

———. *Oeuvres de Turgot*. In *Collection des principaux economistes*. Vols. 3–4. Paris: Guillaumin, 1844.

Vair, Guillaume du. *Les Oeuvres du Sr. du Vair au Parlement de Provence*. Rouen: Jean Osmont, 1614.

Voltaire, François-Marie Arouet de. *Correspondance de Voltaire*. Edited by Theodore Besterman. Tours: Gallimard, 1980.

———. *Le Siècle de Louis XIV*. Paris: Garner-Flammarion, 1966.

———. *Letters on England*. Translated by Leonard Tancock. London: Penguin Books, 1980.

———. *Oeuvres complètes de Voltaire, nouvelle edition*. Paris: Garnier frères, 1877–85.

———. *Voltaire's Notebooks*. Edited by Theodore Besterman. Geneva: Institut et musée Voltaire, 1952.

SECONDARY SOURCES

Adkins, G. Matthew. "The Montmor Discourse: Science and the Ideology of Stability in Old Régime France." *Journal of the Historical Society* 5, no. 1 (March 2005): 1–28.

———. "Reconsidering Political Experience: New Trends in Interpretations of the French Revolution." *The Canadian Journal of History* 39 (August 2004): 325–29.

———. "When Ideas Matter: The Moral Philosophy of Fontenelle." *Journal of the History of Ideas* 61, no. 3 (July 2000): 433–52.

Alengry, Franck. *Condorcet: Guide de la Révolution Française*. Paris: V. Giard & E. Briere, 1904.

Anstey, Roger. "Capitalism and Slavery: A Critique." *Economic History Review* 21, 2nd ser. (1968): 307–20.

Badinter, Elizabeth, and Robert Badinter. *Condorcet: Un intellectuel en politique*. Paris: Fayard, 1988.

Baker, Keith Michael. *Condorcet: From Natural Philosophy to Social Mathematics*. Chicago: University of Chicago Press, 1975.

———. *Inventing the French Revolution: Essays on French Political Culture in the Eighteenth Century*. Cambridge: Cambridge University Press, 1990.

———. "Les débuts de Condorcet au secrétariat de l'Académie royale des sciences (1773–1776)." *Revue d'histoire des sciences* 20 (1967): 229–80.

Bates, David W. *Enlightenment Aberrations: Error and Revolution in France*. Ithaca, NY: Cornell University Press, 2002.

Behrens, C. B. A. *Society, Government and the Enlightenment: The Experiences of Eighteenth-Century France and Prussia*. Thames and Hudson, New York: Harper and Row, 1985.

Beik, William. *Absolutism and Society in Seventeenth-Century France*. Cambridge: Cambridge University Press, 1985.

Bell, David. *The Cult of the Nation in France: Inventing Nationalism, 1680–1800*. Cambridge, MA: Harvard University Press, 2001.

Bender, John, and Michael Marrinan, eds. *Regimes of Description: In the Archive of the Eighteenth Century*. Palo Alto: Stanford University Press, 2005.

Bender, Thomas, ed. *The Antislavery Debate: Capitalism and Abolitionism as a Problem in Historical Interpretation*. Berkeley: University of California Press, 1992.

Bénichou, Paul. *Morales du Grand Siècle*. Paris: Gallimard, 1948.

Bennett, Jane. "Powers of the Hoard: Further Notes on Material Agency." In *Animal, Vegetable, Mineral: Ethics and Objects*, edited by Jeffrey Jerome Cohen, 237–69. Washington, DC: Oliphaunt Books, 2012.

Bigot, Henri. *Les idées de Condorcet sur l'instruction publique*. Thèse, Poitiers: Lévrier, 1912.

Bigourdan, Guillaume. "Les premières réunions savantes de Paris au XVIIe siècle—L'Académie de Montmor." *Mémoires et histoires de l'Académie des Sciences* (January 1917): 159–62, 216–18.

———. *Les Premières Sociétés Savantes de Paris et les Origines de l'Académie des Sciences*. Paris: Gauthier-Villars, 1919.

Birn, Raymond. *Royal Censorship of Books in Eighteenth-Century France*. Stanford, CA: Stanford University Press, 2012.

Bourdieu, Pierre. *Outline of a Theory of Practice*. Translated by Richard Nice. Cambridge: Cambridge University Press, 1977.

Bouton, Cynthia. *The Flour War: Gender, Class, and Community in Late Ancien Régime French Society*. University Park: Pennsylvania State University Press, 1993.

Briggs, Robert. "The Académie Royale des Sciences and the Pursuit of Utility." *Past and Present* 131 (1991): 48–56.

Brooke, Christopher. *Philosophic Pride: Stoicism and Political Thought from Lipsius to Rousseau*. Princeton, NJ: Princeton University Press, 2012.

Brown, Harcourt. *Scientific Organizations in Seventeenth Century France (1620–1680)*. New York: Russell & Russell, 1934.

Bryant, Levi, et al., eds. *The Speculative Turn: Continental Materialism and Realism*. Melbourne, Australia: re.press, 2011.

Burke, Peter. *The Fabrication of Louis XIV*. New Haven, CT: Yale University Press, 1992.

Cahen, Léon. *Condorcet et la Révolution Française*. Paris: Alcan, 1904.

Campbell, P. *Power and Politics in Old Regime France, 1720–1745*. London: Routledge, 199[illegible].

Carré, J.-R. *La Philosophie de Fontenelle, ou le sourire de la raison*. Paris, 1932. Reprint, Geneva: Slatkine Reprints, 1970.

Cassirer, Ernst. *The Philosophy of the Enlightenment*. Translated by Fritz C. A. Koelln and James P. Pettegrove. Princeton, NJ: Princeton University Press, 1951.

Censer, Jack. "Social Twists and Linguistic Turns: Revolutionary Historiography a Decade after the Bicentennial." *French Historical Studies* 22, no. 1 (1999): 139–67.

Chartier, Roger. *The Cultural Origins of the French Revolution*. Translated by Lydia G. Cochrane. Durham, NC: Duke University Press, 1991.

Clarke, J. A. "Abbé Jean-Paul Bignon, 'Moderator of the Academies,' and Royal Librarian." *French Historical Studies* 8, no. 2 (1977): 213–35.

Coutel, Charles. *À l'école du Condorcet: Contre l'Orléanisme des Esprits*. Paris: Elipses, 1996.

Crépel, Pierre, and Christian Gilain. *Condorcet: Mathématicien, économiste, philosophe, homme politique*. Paris: Minerve, 1989.
Dakin, Douglas. *Turgot and the Ancien Régime in France*. 1939. Reprint, New York: Octagon Books, 1972.
Darnton, Robert. *The Forbidden Best-Sellers of Pre-Revolutionary France*. New York: Norton, 1996.
———. *The Literary Underground of the Old Regime*. Cambridge, MA: Harvard University Press, 1982.
———. *Mesmerism and the End of the Enlightenment in France*. Cambridge, MA: Harvard University Press, 1968.
Davis, David Brion. *The Problem of Slavery in the Age of Revolution*. Ithaca, NY: Cornell University Press, 1975.
———. "Reflections on Abolitionism and Ideological Hegemony." *American Historical Review* 92, no. 4 (October 1987): 797–812.
Dear, Peter. *Discipline and Experience: The Mathematical Way in the Scientific Revolution*. Chicago: University of Chicago Press, 1995.
———. *Mersenne and the Learning of the Schools*. Ithaca, NY: Cornell University Press, 1988.
Delorme, Suzanne. "Un Cartésian Ami de Gassendi: Henri-Louis Habert de Montmor." *Revue d'Histoire des Sciences* 27, no. 1 (1974): 20–31.
Dewald, Jonathan. "Crisis, Chronology, and the Shape of European Social History." *American Historical Review* 113, no. 4 (October 2008): 1031–52.
Dick, S. J. *Plurality of Worlds: The Origins of the Extraterrestrial Life Debate from Democritus to Kant*. Cambridge: Cambridge University Press, 1982.
Doeuff, Michèle le. "Bacon chez les grands au siècle de Louis XIII." In *Francis Bacon: Terminologia e Fortuna nel XVII secolo Lessico Intellettuale Europeo XXXIII*, edited by Marta Fattori, 155–78. Rome: Edizionо dell'Ateneo, 1984.
Dorigny, Marcel, and Bernard Gainot. *La Société des Amis des Noirs, 1788–1799: Contribution à l'histoire de l'abolition de l'esclavage*. Paris: Éditions Unesco, 1998.
Doyle, William. *The Oxford History of the French Revolution*. Oxford: Oxford University Press, 1990.
Edelstein, Dan. *The Enlightenment: A Genealogy*. Chicago: University of Chicago Press, 2010.
Evans, Robert. *Jonson, Lipsius, and the Politics of Renaissance Stoicism*. Wakefield: Longwood Academic, 1992.
Fisher, H. A. L. *History of Europe*. New York: Houghton Mifflin, 1935.
Fitzhugh, Michael, and William Leckie Jr. "Agency, Postmodernism, and the Causes of Change." *History and Theory* 40, no. 4 (December 2001): 59–81.
Fitzsimmons, Michael P. *The Night the Old Regime Ended: August 4, 1789, and the French Revolution*. University Park: Pennsylvania State University Press, 2003.
Flourens, Pierre. *Fontenelle, ou de la philosophie moderne relativement aux sciences physiques*. Paris, 1847. Reprint, Geneva: Slatkine Reprints, 1971.
Foucault, Michel. *The Order of Things: An Archaeology of the Human Sciences*. New York: Random House, 1970.
Francisque, Vial. *Condorcet et l'éducation démocratique*. Paris, 1906. Reprint, Geneva: Slatkine Reprints, 1970.
Furet, François. *Interpreting the French Revolution*. Cambridge: Cambridge University Press, 1981.
Gay, Peter. *The Enlightenment: An Interpretation, vol. II: The Science of Freedom*. New York: Knopf, 1969.
———. *The Party of Humanity: Essays in the French Enlightenment*. New York: W. W. Norton, 1954.
———. *Voltaire's Politics: The Poet as Realist*. Princeton, NJ: Princeton University Press, 1959.
Geertz, Clifford. *The Interpretation of Cultures*. New York: Basic Books, 1973.
Gillispie, Charles Coulston, ed. *Dictionary of Scientific Biography*. Vol. 9. New York: Charles Scribner's Sons, 1981.

———. *Science and Polity in France at the End of the Old Regime*. Princeton, NJ: Princeton University Press, 1980.

Goldgar, Ann. *Impolite Learning: Conduct and Community in the Republic of Letters, 1680–1750*. New Haven, CT: Yale University Press, 1995.

Goodman, Dena. *Criticism in Action: Enlightenment Experiments in Political Writing*. Ithaca, NY: Cornell University Press, 1989.

———. *The Republic of Letters: A Cultural History of the French Enlightenment*. Ithaca, NY: Cornell University Press, 1994.

Gordon, Daniel, ed. *Citizens without Sovereignty: Equality and Sociability in French Thought, 1670–1789*. Princeton, NJ: Princeton University Press, 1994.

———. *Postmodernism and the Enlightenment: New Perspectives in Eighteenth-Century French Intellectual History*. New York: Routledge, 2002.

Gould, Philip. *Barbaric Traffic: Commerce and Antislavery in the 18th Century Atlantic World*. Cambridge, MA: Harvard University Press, 2003.

Granger, Gilles-Gaston. *La Mathématique sociale du Marquis de Condorcet*. Paris: Odile Jacob, 1989.

Green, Peter. *Alexander to Actium: The Historical Evolution of the Hellenistic Age*. Berkeley: University of California Press, 1990.

Grégoire, François. *Fontenelle: Une philosophie désabusée*. Nancy: J. Vrin, 1947.

Gruder, Vivian. "Whither Revisionism? Political Perspectives on the Ancien Régime." *French Historical Studies* 20 (1997): 245–85.

Guilloton, Vincent. "Autour de la Relation de Voyage de Samuel Sorbière en Angleterre 1663–1664." *Smith College Studies in Modern Languages* 11, no. 4 (July 1930).

Habermas, Jürgen. *The Structural Transformation of the Public Sphere: An Inquiry into a Category of Bourgeois Society*. Translated by Thomas Burger. Cambridge, MA: MIT Press, 1989.

Hahn, Roger. *The Anatomy of a Scientific Institution: The Paris Academy of Sciences, 1666–1803*. Berkeley: University of California Press, 1971.

Hansen, Millard. "Condorcet's Liberal Philosophy of Education." *South Atlantic Quarterly* 36 (1937): 385–99.

Hanson, Paul R. *The Jacobin Republic under Fire: The Federalist Revolt in the French Revolution*. University Park: Pennsylvania State University Press, 2003.

Hardman, John. *Louis XVI*. London: Arnold, 2000.

Havens, George. *Voltaire's Marginalia on the Pages of Rousseau: A Comparative Study of Ideas*. Columbus: Ohio State University, 1933.

Hazard, Paul. *La crise de la conscience europèene (1680–1715)*. 3 vols. Paris: Boivin et Cie, 1935.

Heller, Henry. *The Bourgeois Revolution in France, 1789–1815*. New York and Oxford: Berghahn Books, 2006.

Hirschman, Albert O. *The Passions and the Interests: Political Arguments for the Capitalism before its Triumph*. Princeton, NJ: Princeton University Press, 1977.

Horkheimer, Max. "Reason Against Itself: Some Remarks on Enlightenment." In *What Is Enlightenment? Eighteenth-Century Answers and Twentieth-Century Questions*, edited by James Schmidt, 359–67. Berkeley: University of California Press, 1996.

Horkheimer, Max, and Theodor W. Adorno. *Dialektik der Aufklärung*. New York: Social Studies Association, 1944.

Hunt, Lynn. *The Family Romance of the French Revolution*. Berkeley: University of California Press, 1992.

———. *Politics, Culture, and Class in the French Revolution*. Berkeley: University of California Press, 1984.

Huppert, George. *The Style of Paris: Renaissance Origins of the French Enlightenment* Bloomington: Indiana University Press, 1999.

Israel, Jonathan. *Enlightenment Contested: Philosophy, Modernity, and the Emancipation of Man, 1670–1752*. New York: Oxford University Press, 2006.

———. *Radical Enlightenment: Philosophy and the Making of Modernity, 1650–1750*. Oxford: Oxford University Press, 2001.

Jennings, Lawrence C. *French Anti-Slavery: The Movement for the Abolition of Slavery in France, 1802–1848*. Cambridge: Cambridge University Press, 2000.

Kahan, Alan. *Aristocratic Liberalism: The Social and Political Thought of Jacob Burckhardt, John Stuart Mill, and Alexis de Tocqueville*. New York: Oxford Press, 1992.

Kaiser, Thomas, and Dale Van Kley, eds. *From Deficit to Deluge: The Origins of the French Revolution*. Palo Alto: Stanford University Press, 2010.

Kapp, Volker. "Les qualitiés scientifique et le prestige social des sciences dans les éloges académique de Fontenelle." In *Fontenelle, actes du colloque tenu à Rouen du 6 au 10 octobre 1987*, edited by Alain Niderst, 435–40. Paris: Presses Universitaires de Frances, 1989.

Keohane, Nannerl O. *Philosophy and the State in France: The Renaissance to the Enlightenment*. Princeton, NJ: Princeton University Press, n.d.

Kintzler, Catherine. *Condorcet: l'Instruction publique et la naissance du citoyen*. Paris: Le Sycomore, 1984.

Kleinbaum, Abby. *Jean Jacques Dortous de Mairain (1678–1771): A study of an Enlightenment Scientist*. Columbia University, PhD dissertation, 1970.

Kuhn, Thomas. *The Structure of Scientific Revolutions*. Chicago: University of Chicago Press, 1962.

Kwass, Michael. *Privilege and the Politics of Taxation in Eighteenth-Century France*. Cambridge: Cambridge University Press, 2000.

LaCapra, Dominick. *Rethinking Intellectual History: Texts, Contexts, Language*. Ithaca, NY: Cornell University Press, 1983.

Latour, Bruno. *Reassembling the Social: An Introduction to Actor-Network Theory*. Oxford: Oxford University Press, 2005.

Lennon, Thomas. *The Battle of the Gods and Giants: The Legacies of Descartes and Gassendi, 1655–1715*. Princeton, NJ: Princeton University Press, 1993.

Lenoble, Robert. *Mersenne ou la Naissance du Mécanisme*. Paris: J. Vrin, 1943.

Levi, Anthony. *French Moralists: The Theory of the Passions, 1585–1649*. Oxford: Clarendon Press, 1964.

Linton, Marisa. *The Politics of Virtue in Enlightenment France*. Houndmills, UK: Palgrave, 2001.

Mah, Harold. "Phantasies of the Public Sphere: Rethinking the Habermas of Historians." *Journal of Modern History* 72 (March 2000): 153–82.

Mandrou, Robert. *From Humanism to Science, 1480–1700*. New York: Penguin Books, 1978.

Manuel, Frank. *The Prophets of Paris*. Cambridge: Cambridge University Press, 1962.

Marsak, Leonard. *Bernard de Fontenelle: The Idea of Science in the French Enlightenment*. Philadelphia: The American Philosophical Society, 1959.

Martin, Kingsley. *The Rise of French Liberal Thought: A Study of Political Ideas from Bayle to Condorcet*. New York: New York University Press, 1929.

Marx, Karl. *The Eighteenth Brumaire of Louis Bonaparte*. New York: International Publishers, 1963.

Maza, Sarah. *Private Lives and Public Affairs: The Causes Célèbres of Pre-Revolutionary France*. Berkeley: University of California Press, 1993.

Mazauric, Simone. *Fontenelle et l'invention de l'histoire des sciences à l'aube des lumières*. Paris: Fayard, 2007.

McClellan, James E., III. *Science Reorganized: Scientific Societies in the Eighteenth Century*. New York: Columbia University Press, 1985.

McLean, Iain. *Condorcet: Foundations of Social Choice and Political Theory*. Brookfield, VT: E. Elgar, 1994.

McMahon, Darrin M. *Enemies of the Enlightenment: The French Counter-Enlightenment and the Making of Modernity*. Oxford: Oxford University Press, 2001.

Meillassoux, Quentin *After Finitude: An Essay on the Necessity of Contingency*. Translated by Ray Brassier. London: Continuum, 2009.

Miller, Peter. *Peiresc's Europe: Learning and Virtue in the Seventeenth Century*. New Haven, CT: Yale University Press, 2000.

Morize, André. "Samuel Sorbière (1610–70)." *Zeitschrift für französische Sprache und Literatur* 33 (1908): 239–44.

Mornet, Daniel. *Les origines intellectuelles de la révolution française, 1715–1787*. Paris: A. Colin, 1933.

Mutha, Sankar. *Enlightenment Against Empire*. Princeton, NJ: Princeton University Press, 2003.

Niderst, Alain. *Fontenelle*. Paris: Plon, 1991.

———. *Fontenelle à la Recherche de lui-même*. Paris: Nizet, 1972.

Nietzsche, Friedrich. *On the Advantage and Disadvantage of History for Life*. 1874. Translated by Peter Preuss. Indianapolis: Hackett, 1980.

Norvins, Jacques Marquet de. *Essais sur la Révolution française*. Paris, 1832.

Oestreich, Gerhard. *Neostoicism and the Early Modern State*. Cambridge: Cambridge University Press, 1983.

Osler, Margaret, ed. *Atoms, Pneuma, and Tranquillity: Epicurean and Stoic Themes in European Thought*. Cambridge: Cambridge University Press, 1991.

Ozouf, Mona. *Festivals and the French Revolution*. Translated by Alan Sheridan. Cambridge, MA: Cambridge University Press, 1988.

Palmer, Robert R. *The Improvement of Humanity: Education and the French Revolution*. Princeton, NJ: Princeton University Press, 1985.

Paul, Charles. *Science and Immortality: The* Éloges *of the Paris Academy of Sciences (1699–1791)*. Berkeley: University of California Press, 1980.

Perkins, Wendy. "The Uses of Science: The Montmor Academy, Samuel Sorbière, and Francis Bacon." *Seventeenth-Century French Studies* 7 (1985): 155–62.

Pintard, René. *Le Libertinage Erudit dans la Première Moitié du XVIIe Siècle*. 2 vols. Paris: Boivin, 1943.

Popkin, Richard H. *The History of Scepticism from Erasmus to Spinoza*. Berkeley: University of California Press, 1979.

———, ed. *The Philosophy of the Sixteenth and Seventeenth Centuries*. New York: Free Press, 1966.

Price, Munro. *Preserving the Monarchy*. Cambridge: Cambridge University Press, 1995.

Readings, Bill. *The University in Ruins*. Cambridge, MA: Harvard University Press, 1996.

Rice, Eugene. *The Renaissance Idea of Wisdom*. Cambridge, MA: Harvard University Press, 1958.

Robinet, Jean-François Eugène. *Condorcet, sa vie, son oeuvre*. Paris, 1893.

Roche, Daniel. *Le siècle des lumières en province: Académies et académiciens provinciaux, 1680–1789*. Paris and the Hague: Mouton, 1978.

Rosenfield, Leonora Cohen, ed. *Condorcet Studies*. Atlantic Highlands, NJ: Humanities Press, 1984.

Rothney, John. *The Brittany Affair and the Crisis of the Ancien Régime*. New York: Oxford University Press, 1969.

Rothschild, Emma. *Economic Sentiments: Adam Smith, Condorcet, and the Enlightenment*. Cambridge, MA: Harvard University Press, 2001.

Sarasohn, Lisa. "Nicolas-Claude Fabri de Peiresc and the Patronage of the New Science in the Seventeenth Century." *Isis* 84 (1993): 70–90.

Schapiro, Jacob S. *Condorcet and the Rise of Liberalism*. New York: Harcourt Brace, 1934.

Sepinwall, Alyssa Goldstein. *The Abbé Grégoire and the French Revolution: The Making of Modern Universalism*. Berkeley: University of California Press, 2005.

Sewell, William H., Jr. *Logics of History: Social Theory and Social Transformations*. Chicago: University of Chicago Press, 2005.

———. *Work and Revolution: The Language of Labor from the Old Regime to 1848*. Cambridge: Cambridge University Press, 1980.

Shank, J. B. *The Newton Wars and the Beginning of the Enlightenment*. Chicago: University of Chicago Press, 2008.

Shapin, Steven. *A Social History of Truth: Civility and Science in Seventeenth-Century England*. Chicago: University of Chicago Press, 1994.

Shelford, April. *Transforming the Republic of Letters: Pierre-Daniel Huet and European Intellectual Life, 1650–1720.* Rochester, NY: Rochester University Press, 2007.

Shovlin, John. *The Political Economy of Virtue: Luxury, Patriotism, and the Origins of the French Revolution.* Ithaca, NY: Cornell University Press, 2006.

Smith, Jay M. "Between *Discourse* and *Experience*: Agency and Ideas in the French Pre-Revolution." *History and Theory* 40, no. 4 (December 2001): 116–42.

———. *Nobility Reimagined: The Patriotic Nation in Eighteenth-Century France.* Ithaca, NY: Cornell University Press, 2005.

Spink, John Stephenson. *French Free-Thought from Gassendi to Voltaire.* London: Athlone Press, 1960.

Stone, Bailey. *Reinterpreting the French Revolution: A Global-Historical Perspective.* Cambridge: Cambridge University Press, 2002.

Stroup, Alice. *A Company of Scientists: Botany, Patronage and Community at the Seventeenth-Century Royal Academy of Sciences.* Berkeley: University of California Press, 1990.

Sturdy, David. *Science and Social Status: The Members of the Académie des Sciences, 1666–1750.* Woodbridge: Boydell Press, 1995.

Sutton, Geoffrey. *Science for a Polite Society: Gender, Culture, and the Demonstration of the Enlightenment.* Boulder: University of Colorado Press, 1995.

Swann, Julian. "Politics and the State in Eighteenth-Century France." In *The Eighteenth Century: Europe 1688–1815*, edited by Timothy C. W. Blanning, 11–51. Oxford: Oxford University Press, 2000.

Tackett, Timothy. *Becoming a Revolutionary: The Deputies of the French National Assembly and the Emergence of a Revolutionary Culture (1789–1790).* Princeton, CT: Princeton University Press, 1996.

Taylor, Charles. *Sources of the Self: The Making of the Modern Identity.* Cambridge, MA: Harvard University Press, 1989.

Tocqueville, Alexis de. *The Old Régime and the French Revolution.* New York: Doubleday, 1955.

Van Meerkerk, Edwin, and Theodore E. D. Braun. "From the *Voyage du baron de Gangan* to *Micromégas*: New Documentary Evidence Linking Voltaire's Space Odysseys." *Studies on Voltaire and the Eighteenth Century* 6 (2002): 385–98.

Wade, Ira. *The Intellectual Development of Voltaire.* Princeton, NJ: Princeton University Press, 1969.

Weber, Eugen. *Peasants into Frenchmen: The Modernization of Rural France, 1870–1914.* Stanford, CA: Stanford University Press, 1976.

White, Hayden. *Tropics of Discourse: Essays in Cultural Criticism.* Baltimore: Johns Hopkins University Press, 1978.

Williams, Eric. *Capitalism and Slavery.* Chapel Hill: University of North Carolina Press, 1944.

Wilson, Arthur. *Diderot.* New York: Oxford University Press, 1972.

Wright, Johnson Kent. *A Classical Republican in Eighteenth-Century France: The Political Thought of Mably.* Stanford, CA: Stanford University Press, 1997.

Yates, Julian. "Toward a Theory of Agentive Drift; Or, a Particular Fondness for Oranges circa 1597." *Parallax* 8, no. 1 (2002): 47–58.

Zinsser, Judith, ed. *Men, Women, and the Birthing of Modern Science.* Dekalb: Northern Illinois University Press, 2005.

Index

About the Author

G. Matthew Adkins earned his PhD in European history from the University of North Carolina at Chapel Hill. He currently teaches European history at Columbus State Community College in Columbus, Ohio. Previously, he taught at the University of Dayton, Miami University, Antioch College, and the City University of New York. In another life he played guitar for an indie punk band, turned down a gig as a CIA analyst, and moved to Brooklyn for two years before finding himself back in Ohio.

www.ingramcontent.com/pod-product-compliance
Lightning Source LLC
LaVergne TN
LVHW102231300926
849160LV00006B/92
* 9 7 8 1 6 4 4 5 3 0 6 4 1 *